Outdoor Lighting for Pedestrians

Outdoor Lighting for Pedestrians shows how outdoor lighting is important for pedestrians' safety, personal security, and comfort, with major impacts on street, path, and park aesthetics and neighborhood sense of place. Providing clear, basic technical background (accessible to non-engineers), the book focuses especially on planning and policy concerns. It covers the fundamentals of lighting technology; benefits, costs, and possible adverse impacts of lighting enhancements; traditional and innovative approaches; planning and policy documents and practices; aesthetics and placemaking; and technology trends in lighting design. This book is aimed primarily at practicing transportation planners and engineers, generalist urban planners, safety advocates and researchers, and university students. However, lighting designers and other professionals will also find it useful. It considers how lighting can be coordinated with other potential improvements to enhance the pedestrian environment for better walkability.

Frank Markowitz is a highly experienced transportation planner and pedestrian safety specialist. His career has been split between public agencies and consulting firms. From 2000 to 2018, he managed pedestrian safety and other planning programs for the San Francisco Municipal Transportation Agency (SFMTA). There he co-led a major federally funded research project evaluating innovative pedestrian safety measures, along with the University of California at Berkeley. His SFMTA projects included consideration of street lighting conditions and improvement concepts, as well as transportation planning and engineering review for large land development projects, several encompassing entire new mixed-use neighborhoods.

Frank initiated and presented two webinars that served as the basis for this book, for the University of North Carolina's Pedestrian & Bicycle Information Center and the Association of Pedestrian & Bicycle Professionals. He was an appointed member of the Transportation Research Board (TRB) Pedestrian Committee and the San Mateo County Bicycle & Pedestrian Advisory Committee. He led two national technical committees of the Institute of Transportation Engineers (ITE). He also prepared over 25 conference presentations, papers, and articles for organizations such as ITE, TRB, the Urban Land Institute, and the American Planning Association. Early in his career, he worked as a research associate for a behavioral sciences research institute and a data analyst for a metropolitan county public health agency. He holds master's degrees in urban planning and health science.

Outdoor Lighting for Pedestrians

A Guide for Safe and Walkable Places

Frank Markowitz

Routledge
Taylor & Francis Group

NEW YORK AND LONDON

Cover image: Courtesy of Buffalo Bayou Partnership

First published 2022
by Routledge
605 Third Avenue, New York, NY 10158

and by Routledge
4 Park Square, Milton Park, Abingdon, Oxon, OX14 4RN

Routledge is an imprint of the Taylor & Francis Group, an informa business

Library of Congress Cataloging-in-Publication Data
A catalog record for this title has been requested

ISBN: 9780367711962 (hbk)
ISBN: 9780367711955 (pbk)
ISBN: 9781003149750 (ebk)

DOI: 10.4324/9781003149750

Typeset in Avenir
by codeMantra

To Nellie, for constantly brightening my days and offering perceptive suggestions.

To Andie Hongying, who made us glow with pride and affection.

And to my parents, for showing the path.

Contents

Figures

Tables

Contributor Biographies

Contributor and Technical Reviewer: Leni Schwendinger

Technical Reviewers: Paul Lutkevich, James Le, and Nancy Clanton

Leni Schwendinger is a published, award-winning authority on issues of city lighting, with more than 20 years of worldwide experience, creating illuminated environments. This work is shared through Leni's public speaking and envisioning engagements, including the "NightSeeing™, Navigate Your Luminous City" program. The Leni Schwendinger Light Projects design projects can be experienced at sites such as parks, subways, and bridges. She is directing a startup, the International Nighttime Design Initiative, establishing an interdisciplinary profession. Projects for the Initiative include Smart Lighting Guidance for New York State and developing innovative pilots with New Urban Mobility, a think-tank. Leni is a Visiting Research Fellow at the London School of Economics and a Design Trust for Public Space Fellow (New York City).

Preface

The genesis of this book was my experience as a transportation planner grappling with pedestrian injuries and fatalities that were frequently "front page news" in San Francisco. In the newly created post of Pedestrian Program Manager for the Municipal Transportation Agency, starting in 2000, I worked with dedicated engineers, planners, and public health professionals to make the pedestrian environment safer, more comfortable, and accessible. We understood that the after-dark environment was a critical factor in safe walking, and lighting conditions played a prominent role in that nighttime world.

However, the outdoor lighting experts for the City and County of San Francisco were based in different agencies. Their standards and tools initially seemed mysterious. While traffic engineering colleagues at my agency had some knowledge of roadway lighting, I found no accessible resources to help someone without a background in lighting understand the technical fundamentals, standards, procedures, and applications of the field.

In preparing to deliver two well-received 2018 webinars for transportation specialists on the topic of this *Guide*, I developed a greater appreciation for the role of outdoor lighting in pedestrian safety and walkability. This underscored the importance of lighting, the rapid changes in technology, and the interest of a broad range of stakeholders in the topic.

The primary aim of this book is to help those who are not lighting specialists understand the potential benefits and challenges of improving lighting, so they can communicate and collaborate effectively with lighting experts. This *Guide* should be useful to a range of professionals and advocates, not just the transportation planner/engineer grappling with the problem that the majority of pedestrian fatalities occur after dark. An urban planner or economic development specialist may delight in the magic of ornamental, futuristic, or artistic lighting on a vacation trip and wonder about the possibility of initiating a similar project back home. An analyst for a police department listens to the complaints of a business group about nighttime crime in a dark commercial strip, questioning what the true value of lighting could be to diminish crime. Citizen advocates for dark skies or pedestrian amenities seek information to support their efforts. It is hoped that all such readers hoping to improve pedestrian safety and walkability can benefit from the information provided on basic lighting technology, applications, trends, and areas for further research.

Acknowledgments

This book would not have been possible without the assistance of numerous experts, friends, and family. The author gratefully acknowledges the support of many individuals (and asks for forgiveness for any omissions).

Leni Schwendinger, Creative Director and Leader of the International Nighttime Design Initiative, provided extensive review, corrections, and suggestions. Her knowledge of the art and science of lighting was invaluable.

Paul Lutkevich, Technical Director and Vice President for WSP engineering consultants, offered many helpful review notes, drawing on his decades of experience and research on roadway lighting projects. Mr. Lutkevich has been involved in landmark standards and professional guidance efforts. James Le, Senior Engineer for the Seattle Department of Transportation, also provided insightful comments and informative graphics. Mr. Le drew on his experience on innovative lighting projects for Seattle. Nancy Clanton, Founder and CEO of Clanton & Associates, a lighting design and engineering firm, reviewed the outline for this book and suggested key resources. Her background, including chairing the Illuminating Engineering Society (IES) Lighting of Public Spaces Committee, was extremely beneficial.

Dan Gelinne, Senior Research Associate for the University of North Carolina's Highway Safety Research Center, organized, moderated, and helped shape the webinar that served as the basis for this book. Dr. Ronald Gibbons, a presenter for that webinar, furnished important information for this book. His leadership in researching roadway lighting as Associate Professor at Virginia Tech (in the VT Transportation Institute and the School of Architecture + Design) provided a source of inspiration. (James Le also presented in that Pedestrian & Bicycle Information Center webinar.) Nicole Hathaway, Senior Development Engineer and Communications Director for the California Lighting Technology Center (at the University of California, Davis), provided additional insights and participated in an earlier Association of Pedestrian & Bicycle Professionals webinar on the topic.

Adriana Lasso Harrier provided expert research and editorial assistance. She patiently obtained images, checked facts and citations, offered suggestions, and reviewed manuscript content. The Routledge/Taylor & Francis editorial team championed and helped shape this book, including Senior Editor Kathryn Schell, Editorial Assistant Sean Speers, Senior Production Editor Nick Craggs, and Production Project Manager Assunta Petrone of codeMantra. Input and ideas were provided by many, including Michael Kato, Associate Public Works Engineer for the City of San Mateo; Brian Liebel, Director of Standards for the IES; Nick Moser, Account Manager for One Hat One Hand design group; and many former SFMTA colleagues, including Adam Smith, Nick Carr, Danielle Harris, Sarah Jones, and Bridget Smith.

My wife Nancy offered incisive ideas, good humor, and patient encouragement throughout the process.

Chapter 1

Introduction

1.1 Why Is This Book Needed?

Lighting is increasingly recognized as a critical factor in the quality or "walkability" of the after-dark pedestrian environment. Pedestrian safety, personal security, and comfort are essential benefits of effective lighting. However, special lighting features and artistic lighting can also improve the aesthetics and unique identity of neighborhoods, tourist zones, and other areas. Lighting enhancements for pedestrians can boost the local economy. Good lighting helps many take advantage of evening and nighttime activities and exercise while enjoying the special after-dark atmosphere.

Lighting also can impose significant costs and potential adverse impacts, although there are measures to mitigate these impacts. Street lighting is a major contributor to municipal energy usage and budgets. Possible health impacts of artificial lighting, especially some light-emitting diodes (LEDs), have attracted the attention of multiple organizations. Unwanted light raises concerns by astronomers about sky glow, by residents and farmers about light "trespassing" on their properties, and by travelers about glare. Luminaires (light fixtures) and poles themselves can raise aesthetic issues and contribute to the hazards of fixed objects adjacent to the roadway.

A general reference is needed for the transportation planner/engineer and others who are not lighting specialists. Lighting technical references are typically aimed at the lighting designer or engineer (often with a background in lighting or architectural engineering, electrical engineering, traffic engineering, or a similar field). Lighting is often overlooked in policy considerations of pedestrian safety and walkability measures, in part because municipal responsibility for street lighting may be in a separate local agency with limited communication with other departments. Lighting needs of pedestrians need to be considered prominently along with those for drivers.

Lighting for pedestrians is an increasingly critical topic because of growing attention to pedestrian safety, dramatic changes in lighting technology, and new interest from professional and activist groups.

DOI: 10.4324/9781003149750-1

This book bridges that gap. It presents technical fundamentals in a manner understandable to those with no technical background in lighting. The *Guide* also includes extensive material on policy issues and case studies to provide a comprehensive, single reference on the most important information on this topic for transportation planners and engineers, safety researchers and advocates, and urban planners and designers, as well as lighting specialists.

This *Guide* comes at an important juncture with a multitude of challenging issues and new directions in lighting. The most important factors include:

1. Increased attention to pedestrian safety and the pedestrian environment
2. New technologies that provide greater control over lighting effects and impacts
3. The involvement of a broad range of stakeholders in lighting issues

1.1.1 Pedestrian Safety and the Pedestrian Environment

Pedestrian safety is a major and growing public health concern. U.S. pedestrian fatalities from motor vehicles hit a 29-year high in 2019, at over 6,500.[1] Fatalities continued to rise during 2020, and with the drop in traffic volumes with the COVID-19 pandemic, the fatality rate per vehicle mile traveled increased 21 percent from 2019.[2] Increased speeding on less congested roads was a likely contributor. Pedestrian fatalities make up 17 percent of national traffic deaths. Nighttime pedestrian deaths increased 67 percent between 2009 and 2018, compared to only 16 percent during daytime. Globally, pedestrians and cyclists comprise 26 percent of traffic fatalities.[3] Pedestrians are 1.5 times more likely than passenger vehicle occupants to be killed in a car crash on each trip.[4] An additional 137,000 estimated pedestrians were treated in hospitals for nonfatal crash-related injuries in 2017.[5]

About three-quarters of pedestrian fatalities in the U.S. occur after dark. Research using 11 years of U.S. Department of Transportation Fatality Analysis Reporting Systems (FARS) data on the distribution of fatal crashes estimated the safety risk to pedestrians to be at least four times higher in darkness than in light.[6]

Lighting improvements can help allay concerns over pedestrian safety and personal security that discourage walking. The Governors Highway Safety Association supported street lighting improvements as a key safety measure (in addition to roadway engineering and enforcement measures, plus decreases in alcohol/drug impairment among drivers and pedestrians).[7] Safety and security concerns can lead to a loss of the significant health and environmental benefits provided by walking or jogging. Walking is "the closest thing we have to a wonder drug," in the words of Dr. Thomas Frieden, former director of the U.S. Centers for Disease Control and Prevention.[8] Besides recognized health benefits of reduced risk for heart disease, diabetes, high blood pressure, and cancer, Harvard Medical School researchers identified several other less-known benefits, like easing joint pain.[9]

Walking is an essential way for residents and visitors to experience cities. As the writer Rebecca Solnit phrased it, "Walkers are 'practitioners of the city,' for the city is made to be walked."[10]

Walkability (including the quality of pedestrian facilities and close proximity of key services) is important to support a high quality of life and attract diverse residents to new homes and travelers to tourist destinations. Homes in walkable neighborhoods command higher prices.[11] One point higher on Walk Score can add nearly one percent to a property's sale price. A survey by real estate firm Redfin found that 56 percent of millennial home buyers, those in their 20s and 30s, said "walkable communities" were a key factor when they looked to buy a home. Good lighting contributes to walkability.

1.1.2 New Technologies

The potential for enhanced lighting dramatically improving the nighttime urban environment was recognized over a century ago at the dawn of the electric age. In 1881, San Jose, California, erected a 237-foot-tall light tower intended to illuminate the entire downtown area.[12] (See Figure 1.1.) The tower failed to sufficiently light the streets, but it led to complaints from farmers about interference with hens laying eggs. (Moonlight towers, originally erected in 1895, remain in Austin, Texas, and are on the National Register of Historic Places.[13]) Around the 1870s and 1880s, electric street lights were installed in parts of New York, London, and Paris. The 1893 World Columbian Exposition's "White City" illuminated a miniature city in Chicago "as if the earth and sky were transformed by the immeasurable wands of colossal magicians" (in the colorful words of a contemporary account).[14] (See Figure 1.2.)

▶ Figure 1.1

1881 San Jose Light Tower. Street scene at night showing electric light tower 250 feet high with 1500 incandescent lights and 12 arc lights in the center of San Jose, California. Courtesy of Wikimedia Commons.

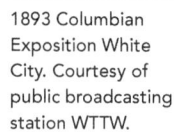

◀ Figure 1.2

1893 Columbian
Exposition White
City. Courtesy of
public broadcasting
station WTTW.

Over the decades since, as electric lighting became commonplace, choices expanded dramatically. There is now an extensive range of pedestrian-serving lighting, from common pole-mounted roadway lights and pedestrian-scale lights to bollards, building-mounted, and ground-level lighting.

LED lighting and "smart lighting" are leading a revolutionary expansion in the ways outdoor lighting can be controlled and used more effectively.[15] "Smart city" sensors to monitor traffic and other activities are often being deployed as part of smart lighting. This has also expanded the interest by researchers in the cost, environmental, and health impacts of these innovations.

1.1.3 Attention from a Broad Range of Stakeholders

Lighting improvements often have the support of safety advocates, businesses, and civic improvement groups. However, lighting technology has also drawn criticism from a broad range of organizations over possible adverse impacts.

The American Medical Association (AMA) in 2016 expressed concern about the possible adverse impacts of "intense blue-rich" LED street lighting, potentially creating a road hazard.[16] The AMA critique listed possible adverse effects of "brighter nighttime lighting" including reduced sleep quantity and quality, daytime sleepiness, impaired daytime functioning, and obesity. Analyses by lighting engineers, discussed in later chapters, have countered these concerns, suggesting that such impacts of well-designed street lighting are minimal and less than that from personal or home electronic devices and indoor lighting.

Obtrusive lighting (sky glow, light trespass, and glare) is a concern for a number of groups and for state legislators. For example, the International Dark-Sky Association, a non-profit organization, "works to help stop light pollution and protect the night skies for present and future generations."[17]

At least 18 states have laws addressing obtrusive lighting.[18] The most common legislation requires on public grounds or right-of-way the installation of shielded light fixtures. Other laws require the use of low-glare or low-wattage lighting, restrict the amount of time that certain lighting can be used, and incorporate Illuminating Engineering Society (IES) guidelines into state regulations.

Residents in some communities have objected in recent years to unwanted, brighter light from certain LED light sources. In Davis, California, complaints led to a public demonstration of different options and installation of replacement lighting following a vote by residents in some neighborhoods.

Such concerns by residents and advocates can be emotional, challenging for proponents of lighting improvements to address. This *Guide* attempts to provide a balanced and objective overview of broad benefits, potential adverse impacts, and measures to mitigate or eliminate such impacts.

1.2 Development of This Book

This *Guide* was developed through a systematic process, tapping the professional expertise of the lighting and transportation engineering communities. It expands and updates the information presented in an October 2018 webinar of the Pedestrian and Bicycle Information Center, housed at the University of North Carolina.[19]

Besides the PBIC webinar, this *Guide* relies primarily on extensive technical references, case studies, conference presentations, and professional and academic experience. The technical references include recommended practices, design guidelines, standards, and manuals produced by lighting and transportation organizations. (These are described more fully in Chapter 2.)

The content was reviewed by lighting experts, with backgrounds in both roadway lighting engineering and lighting design. Technical reviewers were familiar with lighting research and standards, as well as the social and aesthetic aspects of lighting.

1.3 Scope and Structure of This Book

This *Guide* aims to include the most important information for planning, policy, and systems engineering purposes. It should help the transportation professional, urban planner/designer, or safety researcher/advocate communicate with lighting specialists and collaborate on addressing problems of common interest. This *Guide* is not intended to be used for design purposes or for legal advice.

The emphasis is on lighting in the street right-of-way. However, there is a limited treatment of lighting for off-street facilities, such as plazas and pedes-

This Guide takes advantage of input from transportation and lighting design experts, both practitioners and researchers.

trian paths in parks, waterfronts, or similar settings. Such off-street locations are very important, but less often walked for utility trips, and their pedestrian environments vary so greatly that guidance is more difficult.

The initial chapters provide general background. Chapter 2 presents an overview of key technical concepts and terms. It describes key references in the field, as well as legal liability considerations for local governments and other key organizations. Chapters 3 and 4 present the benefits and potential challenges of lighting improvements.

The remainder of the *Guide* focuses on the state of the art and likely future directions in lighting for pedestrians. Chapter 5 describes the standard options available for lighting pedestrian facilities: conventional roadway-scale lights and pedestrian-scale lighting, lighting source technology, and visibility supplements to lighting for improving after-dark safety. Chapter 6 discusses new technologies available now or imminently, especially smart lighting (including adaptive lighting, networked and remotely controlled lighting, and "smart city" applications). Chapter 7 reviews planning and policy considerations as expressed primarily in policy documents like municipal lighting master plans. Chapter 8 discusses how lighting considerations can be integrated with transportation design, operations, and maintenance efforts that affect the pedestrian realm. Chapter 9 addresses placemaking and aesthetic options and considerations. The *Guide* closes in Chapter 10 with an analysis of key trends and measures, discussing the need to "future proof" policy documents and designs to prepare for unpredictable scenarios.

Notes

1. "Pedestrian Traffic Fatalities by State: 2019 Preliminary Data," Governors Highway Safety Association, accessed March 1, 2021, https://www.ghsa.org/resources/Pedestrians20.
2. Richard Retting, *Pedestrian Traffic Fatalities by State: 2020 Preliminary Data and Addendum* (Washington, DC: Governors Highway Safety Association), https://www.ghsa.org/resources/Pedestrians21.
3. World Health Organization, *Global Status Report on Road Safety* (Geneva: WHO, 2018), https://www.who.int/publications/i/item/9789241565684.
4. L.F. Beck, A. M. Dellinger, M.E. O'Neil, "Motor Vehicle Crash Injury Rates by Mode of Travel, United States: Using Exposure-Based Methods to Quantify Differences," *American Journal of Epidemiology* 166, no. 2 (2007): 212–218, doi: 10.1093/aje/kwm064.
5. "WISQARS (Web-based Injury Statistics Query and Reporting System)," Centers for Disease Control and Prevention, accessed October 15, 2019, http://www.cdc.gov/injury/wisqars.
6. John M. Sullivan and Michael J. Flannagan, *Characteristics of Pedestrian Risk in Darkness*, Technical Report: UMTRI 2001–33 (Ann Arbor: University of Michigan Transportation Research Institute, 2001).
7. Governors Highway Safety Association, "Pedestrian Traffic Fatalities."

8. Alexandra Sifferlin, "An Easy Way to Get Enough Exercise: Take a Walk," *Time Online*, August 7, 2012, http://healthland.time.com/2012/08/07/an-easy-way-to-meet-physical-activity-guidelines-take-a-walk/.

9. Harvard Medical School, *Walking for Health: Why This Simple Activity Could Be the Best Health Insurance* (Cambridge, MA: Harvard Health Publications, 2015).

10. Rebecca Solnit, *Wanderlust: A History of Walking* (Brooklyn, NY: Verso Books, 2001).

11. Daniel Goldstein, "How a High 'Walk Score' Boosts Your Home's Value," MarketWatch Website, August 11, 2016, https://www.marketwatch.com/story/how-walk-score-boosts-your-homes-value-2016-08-11.

12. Clyde Arbuckle, *Clyde Arbuckle's History of San José* (San Jose, CA: Memorabilia of San Jose, 1986), 497–498.

13. Mark Oppenheimer, "Austin's Moon Towers, Beyond 'Dazed and Confused,'" *New York Times*, February 13, 2014, https://www.nytimes.com/2014/02/16/travel/austins-moon-towers-beyond-dazed-and-confused.html?_r=0.

14. Charles E. Skinner, "Lighting the World's Columbian Exposition," *Western Pennsylvania History* 17, no. 1 (1934): 13, https://journals.psu.edu/wph/article/viewFile/1666/1514.

15. Felicity Barringer, "New Technology Inspires a Rethinking of Light," *New York Times*, April 24, 2013, https://www.nytimes.com/2013/04/25/business/energy-environment/new-technology-inspires-a-rethinking-of-light.html?partner=rss&emc=rss&smid=tw-nytimes&_r=1&.

16.. American Medical Association Council on Science and Public Health, *Human and Environmental Effects of Light Emitting Diode (LED) Community Lighting, Report 2-A-16* (Chicago: AMA, 2016), https://www.ama-assn.org/sites/ama-assn.org/files/corp/media-browser/public/about-ama/councils/Council%20Reports/council-on-science-public-health/a16-csaph2.pdf.

17. "Light Pollution," International Dark-Sky Association Website, accessed March 2, 2021, https://www.darksky.org/light.

18. "States Shut Out Light Pollution," National Conference of State Legislatures, accessed March 2, 2021, http://www.ncsl.org/research/environment-and-natural-resources/states-shut-out-light-pollution.aspx.

19. Pedestrian and Bicycle Information Center (PBIC) Webinar, "Lighting for Pedestrian Safety and Walkability," October 17, 2018, http://www.pedbikeinfo.org/pdf/Webinar_PBIC_101718.pdf.

Chapter 2

Lighting 101

Technical Fundamentals

2.1 Purpose and Scope of This Chapter

The purpose of this chapter is to provide those who are not lighting specialists sufficient technical and contextual background to address the topic effectively and to communicate clearly with lighting engineers and designers. (Lighting specialists often differentiate between "lighting designers," who typically have broad training in lighting science and art, often with an architectural or other design background, and "lighting engineers," who often have primarily electrical or traffic engineering training. This book refers to all those preparing lighting designs as "designers.") Recommendations for enhanced lighting at specific locations need to account for a host of considerations, including recommended lighting levels, location characteristics, pertinent policies, and existing and planned physical features.

This chapter first discusses basic lighting concepts and terms, relating street lighting to human vision. It then discusses lighting equipment (especially luminaires and mountings) and recommended light levels. Finally, it describes key technical references and legal considerations.

The lighting technology and equipment employed affect the potential benefits and financial costs, and can have other potential adverse impacts on lighting changes (discussed in Chapters 3 and 4). While there are recognized standards, criteria, and best practices, these are continually being refined to reflect new research findings and changes in policies, technology, and the urban environment itself.

2.2 Basic Lighting Concepts and Terms

The effectiveness of outdoor lighting for pedestrian safety and comfort depends on the lighting equipment, characteristics of the pedestrians, and the physical environment. This complicates the tasks of recommending and assessing light levels. While recommendations for light levels are provided by professional organizations, the lighting designer is accorded a great deal of flexibility to respond to unique or complicated situations. Lighting criteria change to adjust to research findings and new measurement tools. For example, measuring and controlling glare is a complicated topic, with differ-

DOI: 10.4324/9781003149750-2

ent glare metrics, measuring devices, and considerations as technology has changed through the years.

2.2.1 Vision and Lighting

The purpose of street lighting is to enhance safety by assisting drivers, bicyclists, and pedestrians in seeing each other and fixed objects in low-light conditions. It also helps pedestrians feel more secure and comfortable, avoid hazards, size up other pedestrians, navigate, and perceive details of buildings and other features. Street lighting can also compensate for the reduction in visibility due to glare from vehicle headlamps.

The effectiveness of street lighting depends on a number of factors. These include the placement and design of the luminaires (lighting fixtures), the street environment (such as the roadway geometrics, pavement reflectivity, and tree canopy), other lighting sources, and the characteristics of drivers and pedestrians (such as age, perceptual impairments, and clothing worn by pedestrians). Some of these factors are complex and often not obvious. For example, the level of glare a driver experiences depends on such variables as the driver's age and the angle of oncoming vehicle headlamps, in turn affected by any median island and other roadway characteristics.

Approximately 90 percent of the information a driver needs comes from visual cues, a stream of sights ranging from static signs to rapidly moving vehicles.[1] The driver's complex visual task is made harder by low light conditions and varying light levels.

Visibility on the roadway is affected by a range of factors, including lighting levels, the color of objects, and glare in the traveler's eyes. It is complicated by the numerous moving and fixed objects a driver or pedestrian views and the different luminance (brightness) levels encountered.

Contrast of objects is critical to perceiving visual cues quickly and accurately. Contrast is either positive (with the primary object brighter than surroundings) or negative (when the primary object is in silhouette against brighter surrounding objects). Under glare-free, uniform background conditions, the contrast that provides a 50 percent chance of detecting an object on the roadway is called the "threshold contrast." How much contrast is needed to perceive an object depends on such factors as the size and color of the object, how long it is viewed, the viewer's age and disabilities, luminance of the road, glare, and ambient lighting. One purpose of roadway lighting is to produce sufficient contrast of objects on the roadway. (This contrast-heightening effect is produced by a combination of contrast of luminances and colors. Color contrast is influenced by spectral content of lighting.)

The impacts of lighting on a particular pedestrian facility (e.g., sidewalk, crosswalk, or separated walkway) and its users depend on a broad number of factors, including lighting levels, roadway features, and the age and clothing of pedestrians.

Glare is an unwanted bright light shining into the eyes of the traveler or observer (like vehicle headlamps) that is hard for the eye to adapt to, thereby interfering with proper vision. Glare occurs when light is scattered within the eye and superimposes a luminous haze on the retina, like a veil. (This phenomenon is often referred to as "veiling luminance.") There are two main types of glare effects. "Disability glare" interferes directly with tasks like driving. "Discomfort glare" refers to the blinking, tears, or pain that is not accompanied by substantially reduced vision.

The eye's sensitivity to light varies for different **wavelengths or frequencies** of light reflected off objects. This is the reason, for example, that fluorescent yellow-green warning signs are more effective than other colors.

Viewing the roadway itself, different **pavement surface types** have varying reflectance characteristics. Roadway surfaces are classified from R1 to R4, based on the reflectivity of the pavement and how bright the pavement appears. R1 on the bright end of the scale includes concrete and asphalt with a higher level of artificial brightener aggregates, while R4 includes very smooth asphalt road surfaces (which appear darker and duller). R1 surfaces reflect light in a more focused manner, while R4 surfaces reflect light more diffusely.

Visual acuity generally is reduced by **age**.[2] Increasing age decreases the eye's sensitivity, especially to blue light. Age-related changes also increase the scatter of light in the eye, leading to more problems with glare. Serious vision impairments (such as macular degeneration, cataracts, and glaucoma) also are more common among the elderly.

2.2.2 Measuring Light Levels and Visibility

Light levels (as perceived by human vision) are measured by photometry, either in the laboratory or field. Light levels are also estimated by software programs. The following are key measurement terms. (See Figure 2.1 for a visual explanation.)

Lumens quantify the total amount of light emitted in all directions by a source. This is analogous to the total volume of water sprayed by a sprinkler.

Candlepower is the luminous intensity or concentration of light emitted in a particular direction, expressed in **candelas (cd).** This brightness measure typically varies at different angles from the light source. It is similar to the force of water sprayed by the sprinkler in one direction.

Illuminance is the density of light falling on a surface. It is expressed as **lux** (lumens per square meter) or **foot-candles** (or "fc," lumens per square foot). It is analogous to the amount of water from a sprinkler falling on a given surface area. It typically measures light directly from a light source.

Luminance is the intensity of light reflected off a surface. It is expressed as candelas per square foot or square meter. It is similar to the force of water bouncing off a given surface area toward an observer. It is closer to a measure of the light the observer usually sees than illuminance, but it is generally more complicated to measure in the field. A background like a roadway surface with

▶ Figure 2.1

Illustration
of Lighting
Terms. Figure
reproduced with
permission from
The Illuminating
Engineering Society
© ANSI/IES RP-8-18.

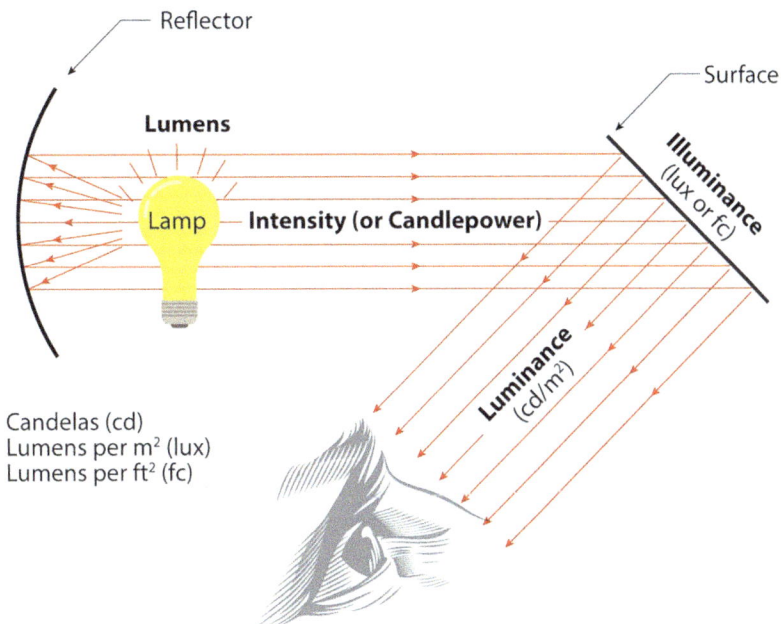

a high luminance requires a higher level of illumination on the pedestrian to improve the walker's visibility to a driver.

Uniformity ratios measure the difference between average and minimum (or maximum versus minimum) illuminance levels. Thus, a high uniformity ratio actually represents less uniformity. Lighting standards and most experts traditionally considered more uniform lighting desirable. However, the City of San Jose presented research evidence in its *Public Streetlight Design Guide* that highly uniform lighting with light-emitting diodes (LEDs) resulted in poorer target detection (especially by older persons), more energy use, and higher glare.[3] Nuanced shadows help with object detection by increasing contrast.[4] The San Jose *Guide* therefore recommended a maximum to minimum illuminance ratio of 8 or more, higher than typical standards for major streets. However, pedestrian reassurance (or feelings of confidence) is enhanced by more uniform lighting of walkways.[5]

Visibility metrics also address glare and object visibility. **Disability glare** is measured by **veiling luminance ratio** or **threshold increment.** Veiling luminance ratio, typically used in North America, relates the luminance of scattered light within the eye (the "veiling" effect) to the luminance of the roadway surface. It is typically worsened by age. Specific age factors have been developed to address this in calculating veiling luminance.[6]

Small target visibility is an alternative metric for visibility of a roadway section based on an array of small targets on the roadway. (See Figure 2.2.) It has the theoretical advantage of being a broader single metric of lighting conditions than others mentioned above. However, it is not generally used in guidelines or standards, and it is more challenging to estimate or measure. It accounts simul-

(a)  (b)

taneously for target luminance, roadway luminance, adaptation level of immediate surroundings, and disability glare. Adaptation level refers to the sensitivity of the eye (at a particular time and in a specific visual field) to changes in light levels of objects. For example, the eye is less sensitive to typical changes like the presence of headlights during daytime.

Relative visual performance (RVP) is an alternative metric for the speed and accuracy of visual processing of a roadway segment.[7] Its value is a function of background luminance, luminance contrast, and the visual size of objects. The potential advantage of RVP, for example for evaluating different crosswalk designs, is that it accounts for factors beyond vertical illuminance. Increasing vertical illuminance does not necessarily improve the visibility of a pedestrian if there is minimal contrast between the pedestrian and the roadway. RVP is mentioned in Chapter 6 regarding tests of bollard lighting.

▲ Figure 2.2

Small Target Visibility. Higher STV in Figure 2.2(b) on the right. Figure reproduced with permission from The Illuminating Engineering Society © ANSI/IES RP-8-18 and photographer James Havard.

2.2.3 Luminaires and Other Equipment

Roadway lighting designers select the best mounting height, luminaire spacing, and lighting output combination to meet light level standards and achieve optimal efficiency. When designing for pedestrians as well as drivers, however, other lighting impacts such as personal security and aesthetics (of both the light characteristics and the luminaires and mountings) are also important. Pedestrians need to be able to view the details of the walkway, other pedestrians, and built environment features. Lighting designers also need to consider such factors as life cycle costs, potential fixed object hazards, and ease of maintenance of equipment.

Street lights are often differentiated by the type of mounting of luminaires. (See Figure 2.3 depicting street light components.) Luminaires are commonly mounted on extensions from poles (davit-style, mast arms, or truss style). They can also be mounted on bollards, atop posts or high masts, on walls, on utility poles, as flood lights, as linear systems (e.g., atop bus shelters), from catenary wires, or in the roadway. For the common pole-mounted system, equipment includes a foundation and base, the pole itself, possibly a projecting arm, and the luminaire. The luminaire includes the light source (lamps or diodes), along

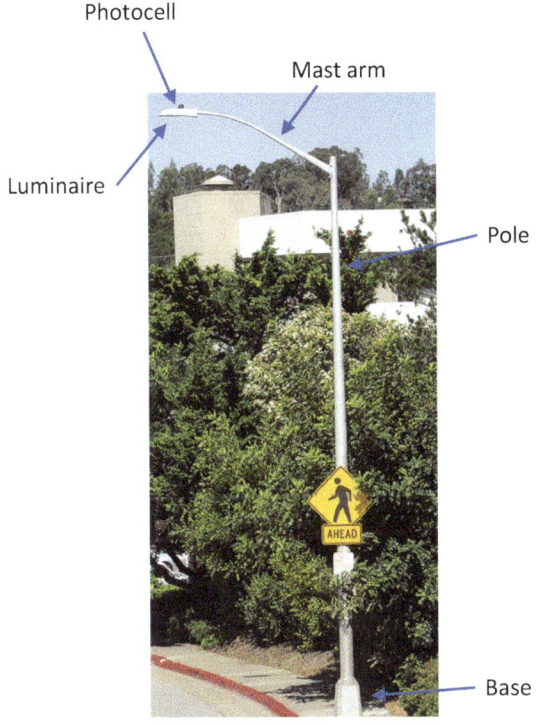

with optics (parts to distribute the light), parts to position and protect lamps, and devices to connect lamps to power. Sometimes luminaires include ballasts (to regulate the current to the lamp) and photocells (to switch on the light after dark). Electrical service connection locations and the extent of trenching and underground conduit to the lighting equipment can be major cost drivers.

Light distribution characteristics can be described by the Luminaire Classification System for Outdoor Luminaires (LCS) and quantified by BUG (Backlight, Uplight, Glare) ratings. (The primary angles for the LCS are shown in Figure 2.4.) The LCS describes lumen output within solid angles in three directions: forward, back, and uplight. The luminaire can also be assigned a BUG rating according to the lumen output in the various Backlight, Uplight, and Glare zones. This BUG rating can be helpful in selecting the proper luminaire to minimize light trespass, sky glow, and glare. However, the BUG rating is a simplified metric for roadway lighting, especially related to glare. The BUG rating should be considered along with the level of ambient lighting. In a zone that already has a high level of ambient lighting (like an active nighttime retail area), a higher BUG rating for a new luminaire is less troublesome than if the same luminaire is installed in an area with little or no ambient light.

BUG ratings are affected by the light source output (lumens) and color temperature (CCT), the optical distribution, power (watts), and lens choice.[8]

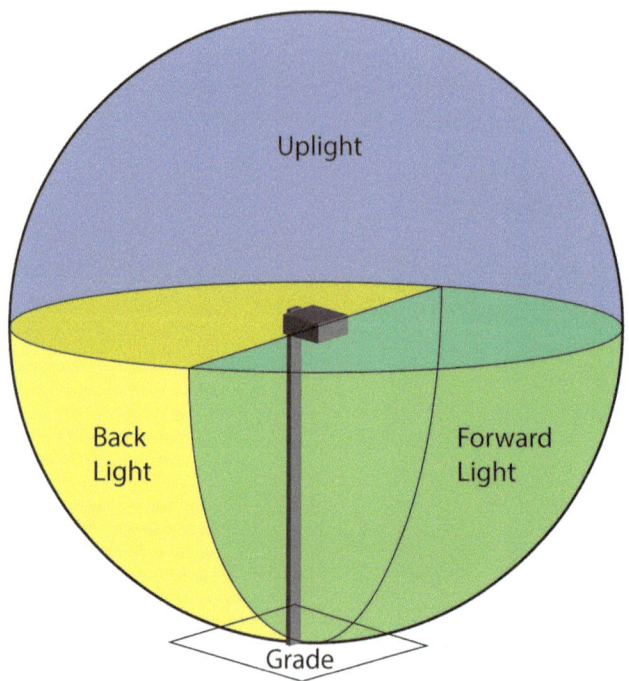

◀ Figure 2.4

Luminaire Classification System: Primary Angles. Figure reproduced with permission from The Illuminating Engineering Society © ANSI/IES RP-8-18.

(a)

(b)

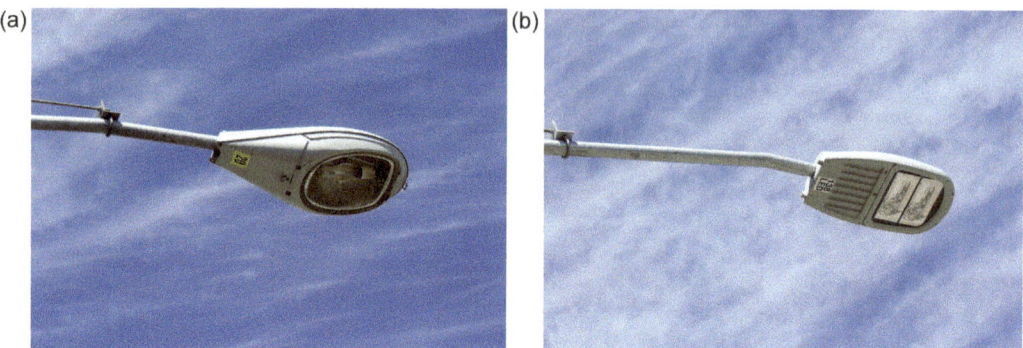

BUG ratings go from 000 to 555. However, 000 is not always ideal. For example, bollards with a BUG rating of 000 may still generate glare complaints.

There are numerous light sources available. New roadway lighting installations routinely use LEDs. (See Figure 2.5.) However, still very common on roadways is some form of high-intensity discharge, such as high-pressure sodium (HPS). Other less common options include low-pressure sodium, compact fluorescent, incandescent, and induction lamps. Further information on light source technology is included in Chapter 5.

▲ Figure 2.5

High-Pressure Sodium (HPS, left photo) and Light Emitting Diode (LED, right photo) Luminaires. Courtesy of *My Ferndale News*.

2.2.4 Light Level Recommended Values

Appropriate light levels for different facilities used by pedestrians have been recommended by professional organizations. The IES, representing lighting engineers and specialists, has provided recommended levels that are commonly accepted in the U.S. The IES typically obtains formal standards review from organizations such as the American National Standards Institute (ANSI).

Different **lighting metrics** are used for different types of facilities. Typically, pavement luminance (reflective brightness) is used to design midblock highway and street lighting. Horizontal and vertical illuminance are used for pedestrian areas. (Horizontal illuminance represents the light falling on the roadway surface such as crosswalk markings. Vertical illuminance represents the light falling on a pedestrian, but measured as if light falls on a vertical plane.) Horizontal illuminance is used for intersections, interchanges, and curved sections of roadways where luminance can be difficult to calculate. Research is ongoing to determine whether semi-cylindrical illuminance is a better measure for pedestrian detection than vertical illuminance on a plane.[9]

Lighting designs are analyzed with a **photometric plan**, a layout of the intersection, or street segment.[10] Typically, the plan shows illuminance levels in foot-candles or lux on a grid of points, plus the location and description of luminaires. Calculations may be provided in a table that allows the comparison of light levels to the standards summarized below.

The IES RP-8 *Recommended Practice* recommended light levels are provided in a set of tables, for multiple locations: mid-block (between intersection), intersections, roundabouts, and walkways. IES RP-8 Table 11-1 provides **mid-block street design lighting criteria**. (Table 2.1 shows the structure of the IES table. Tables with the full set of values are proprietary information. They are available from IES, and maintained as part of the Lighting Library, whose contents are available by subscription and automatically updated.[11]) In addition to the recommended minimum average roadway pavement luminance, average uniformity ratio and maximum uniformity ratio are provided to ensure that luminance is consistent traveling along the roadway. Maximum veiling luminance ratio is provided to help control glare.

Higher luminance levels are recommended for streets with higher traffic and pedestrian volumes. In the table, streets are classified based on traffic volumes, roadway use, and adjacent land uses. "Major streets" are arterials or thoroughfares, primarily serving through traffic, and carrying heavy traffic volumes. "Local streets" primarily provide direct access to residential, commercial, industrial, and other land uses, typically with low traffic volumes. "Collector streets" connect major and local streets, and they generally have moderate volumes, with a balance between serving through traffic and providing land use access.

Pedestrian activity is not quantified for this particular table, but it is based on estimated nighttime or low light volumes. "High pedestrian activity" is typically found in commercial areas. "Medium pedestrian activity" is typical of streets near community facilities, like libraries. "Low pedestrian activity" is generally found in purely residential areas.

While the light level differences based on pedestrian activity are primarily related to pedestrians on the sidewalk, the potential for pedestrians to cross streets mid-block, especially at certain locations, should be considered. Even in cases where mid-block crossing is illegal, like on blocks signalized at both ends, it may be much more attractive to some pedestrians than walking to the nearest intersection. On local and collector blocks, mid-block crossing is usually legal.

Intersection lighting is based on different design factors than mid-block locations since intersections are more complex and potential vehicle and pedestrian conflicts are much greater. Lighting designers need to consider safety, cost, aesthetics, site conditions such as vegetation, and minimizing obtrusive light.

The IES RP-8 Table 12.1 recommends roadway pavement horizontal illuminance levels for intersections with full lighting. (See Table 2.2 for the structure of the IES table. "Full lighting" is intended to illuminate the entire intersection, not just conflict points.) The publication also provides recommended levels for intersections with partial lighting or delineation lighting only. ("Partial lighting"

▼ Table 2.1

IES Recommended Light Levels for Streets (Mid-Block)

Street Classification	Pedestrian Activity Classification	Average Luminance L_{avg} (cd/m^2)	Average Uniformity Ratio L_{avg}/L_{min}	Maximum Uniformity Ratio L_{max}/L_{min}	Maximum Veiling Luminance Ratio $L_{v,max}/L_{avg}$
Major	High	1.2	3.0	5.0	0.3
	Medium				
	Low				
Collector	High				
	Medium				
	Low				
Local	High				
	Medium				
	Low				

Notes:
L_{avg}: Maintained average pavement luminance.
L_{min}, L_{max}: Minimum and maximum pavement luminance.
$L_{v,max}$: Maximum veiling luminance.
Most cell values are not provided, as the purpose is just to show the structure of the table. Complete values available from Illuminating Engineering Society, Recommended Practice 8-18: Design and Maintenance of Roadway and Parking Facility Lighting (New York: IES, 2018), Table 11-1.
Table reproduced with permission from The Illuminating Engineering Society © ANSI/IES RP-8-18.

▼ Table 2.2

IES Recommended Light Levels for Intersections Full Intersection lighting – recommended illuminance (lux/fc)

Functional Classification	Pedestrian Activity Level Classification			Maximum Uniformity Ratio E_{avg}/E_{min}
	High	Medium	Low	
Major/Major	34/3.2	26/2.4	18/1.7	3.0
Major/Collector				
Major/Local				
Collector/Collector				
Collector/Local				
Local/Local				

Notes:
E_{avg}/E_{min}: Maximum Uniformity Ratio (Average Illuminance divided by Minimum Illuminance).
Complete values available from Illuminating Engineering Society, Recommended Practice 8-18. Design and Maintenance of Roadway and Parking Facility Lighting (New York: IES, 2018), Table 12-1.
Table reproduced with permission from The Illuminating Engineering Society © ANSI/IES RP-8-18.

for intersections is intended to illuminate primarily conflict points. "Delineation" or "beacon lighting" is merely intended to indicate the location of an intersection, rather than to illuminate it fully.) The table assumes R2 and R3 pavements (moderate brightness and reflectivity).

Higher illumination levels are recommended in the table for intersections of major streets (typically with higher traffic volumes) and for higher pedestrian volumes. Illuminance is expressed in both lux and footcandles. Also, a uniformity ratio (average illuminance divided by minimum illuminance) is recommended. Veiling luminance criteria are not provided because of the difficulty of calculating for an intersection, but IES RP-8 recommends using luminaires with a more vertical light distribution.

The functional classifications for streets are as described in the previous section for mid-block locations. Specific pedestrian volume ranges are provided to define "pedestrian activity" levels for crossings. "High pedestrian activity" is based on 100 or more pedestrians over the one-hour period with the highest average annual nighttime pedestrian volumes. "Medium pedestrian activity" reflects 11–99 pedestrians per hour, and "low pedestrian activity" is based on 10 or fewer pedestrians per hour.

Illuminance levels are recommended for the box formed by imaginary curb line extensions. Other "conflict areas" adjacent to the intersection should have illumination 50 percent higher than these levels.

Regarding **crosswalks in intersections**, IES RP-8 also refers to a Federal Highway Administration (FHWA) study performed by the Virginia Tech Transportation Institute. It concluded that vertical illumination levels at 1.5 meters (5 feet) high in a midblock crosswalk of 20 lux should be sufficient for drivers to detect pedestrians in rural conditions at a safe stopping distance. Glare and overall lighting conditions (high ambient lighting and roadway lighting for vehicles) may justify higher vertical illumination for pedestrians.

However, the recommendations do not explicitly include the sidewalks immediately adjacent to the crosswalks. The visibility of pedestrians before stepping into a crosswalk is a major safety concern (explained in the later sidebar).

A recent international review of lighting research suggested that the minimum vertical illumination for pedestrian crossings should be 6 lux.[12] This considered both traffic safety and tripping issues.

IES RP-8 Table 12-4 provides recommended illuminances for **roundabouts** that are essentially the same as those for full intersection lighting. However, the Recommended Practice lays out a detailed procedure for estimating horizontal (roadway surface) illuminance in a grid from the splitter islands into the roundabout. It also provides a similar procedure for vertical illuminance intended to cover the visibility of pedestrians. The grid does not include the sidewalk next to crosswalks. In crosswalks, vertical illuminance at a height of 1.5 meters (5 feet), as seen from the drivers' direction, at the same recommended levels as the horizontal illuminance, should also be confirmed.

For **midblock crosswalks**, IES RP-8 bases its recommendations on the FHWA study by the Virginia Tech Transportation Institute mentioned earlier. It recommends maintained average vertical illuminance of 20–40 lux: 20 lux for crossings with low pedestrian crossing volumes, 30 lux for medium pedestrian crossing volumes, and 40 lux for high pedestrian volumes. Vertical illuminance levels should equal or exceed horizontal light levels to help ensure sufficient contrast of the pedestrian against the pavement.

For **walkways/bikeways within the street right-of-way**, IES RP-8 provides recommended light levels separately for high, medium, and low pedestrian activity. Walkways include sidewalks and paths adjacent to the roadway. (Walkways completely separated from the street, such as internal park paths, are not addressed in this document, but are discussed in a separate IES document, LP-2-20, *Lighting Practice: Designing Quality Lighting for People in Outdoor Environments*.[13]) IES RP-8-18 notes that these levels do not address personal security needs, such as in high-crime areas, but these considerations are addressed in a separate IES document, G-1-16, *Security Lighting for People, Property and Critical Infrastructure*.[14]

Minimum horizontal and vertical illuminance values are provided, along with recommended uniformity ratios. Higher rates are given for mixed pedestrian/vehicle areas than for pedestrian-only walkways. (Mixed pedestrian/vehicle areas are primarily for "shared streets," woonerfs, or plazas where pedestrians have priority, but slow vehicles are allowed. These have been rarely used in the U.S., but they are increasing in popularity.) The recommended values are generally lower than corresponding intersection levels. There are no specific glare criteria, however, the publication recommends minimizing glare in the design. To ensure adequate sidewalk lighting at intersections, typically, the roadway portion is designed first for vehicles, then the designer evaluates whether that

Light levels on the roadway and on adjacent sidewalks are critically interrelated in their impacts on pedestrian safety.

Lighting Levels for Crosswalks vs. Sidewalk Approaches

Drivers need to perceive pedestrians approaching crosswalks at night. A pedestrian on a street corner who is significantly darker than the intersection may be hard to observe as he or she enters the crosswalk. The driver may not have time to react to a pedestrian who appears to suddenly "pop" into view. (See Figure 2.6.)

The lighting levels recommended by the IES for sidewalks are slightly lower than corresponding intersection requirements. (For high pedestrian activity, pedestrian-only sidewalks are to have a minimum of 10.0 lux horizontal illuminance at pavement level

and 5.0 lux vertical illuminance at 5 feet, but arterial intersections with high pedestrian activity are to be lit at least 26–34 lux at pavement level, depending on the intersecting streets' classifications.)

The recently released *Solid-State Roadway Lighting Design Guide* focuses on the needs for ensuring that sidewalks receive sufficient lighting while limiting light trespass on adjacent private properties. The *Solid-State Guide* provides an example of a computer-generated lighting grid that includes sidewalks and adjacent properties.

▲ Figure 2.6

Street Corners Darker Than Intersection. Courtesy of Adobe Stock Photos. © freebreath - stock.adobe.com.

When outdoor lighting nearly matches the color rendition of daylight, object contrast is improved and safety enhanced.

lighting will also meet the sidewalk (walkway) requirements. If necessary, supplemental lighting is designed, such as pedestrian-scale lighting.

2.2.5 Color Rendition

The **color rendition** of the light source is important to effective street lighting. Visual performance and pedestrian feelings of security are improved by lighting that shows colors of objects closer to daylight conditions, providing higher contrast. The **Color Rendering Index** (CRI) is a widely used metric for the ability of a light source to match faithfully the color delineation produced by natural lighting. IES *Tech Memo 30* in 2018 proposed replacing the CRI with gamut and fidelity indices.[15] The fidelity index uses 99 color samples, so it is more precise than the CRI, which uses only eight color samples. The gamut index measures the range of intensity or saturation of colors revealed by the light source. A color vector graphic can also be used to show which colors are more saturated by the light source. However, as of publication, CRI was still commonly used in lighting specifications and vendor descriptions. LED lighting generally has a higher CRI than high-pressure sodium and metal halide. (See Figure 2.7 for a comparison of different CRI levels.)

Health impacts of different color content light are also discussed in Chapter 4. To understand these issues, it is helpful to appreciate the spectral characteristics of lighting. Human vision is more sensitive to light of specific wavelengths, depending on overall lighting conditions. In **photopic**, or daylight conditions, when sight relies heavily on cone receptors (concentrated in the center of the retina), eyes are most sensitive to green and yellow light. In **scotopic**, or very low light conditions, when sight relies heavily on rod receptors (concentrated in the periphery of the field of vision), eyes are most sensitive to violet, blue, and blue-green light. Typical nighttime driving or walking lighting conditions are in the intermediate **mesopic** range, in between photopic and scotopic ranges.

Any street light source emits light of multiple wavelengths, at different power levels. **Spectral power distribution** diagrams show the intensity of light emitted at different wavelengths by different light sources.

The **Correlated Color Temperature (CCT)** refers to the color appearance (warmth or coolness) of light emitted by a light source. It allows color references to be standardized and expressed simply as a single number of degrees on the

▲ Figure 2.7

Color Rendering Index. Courtesy of Take Three Lighting.

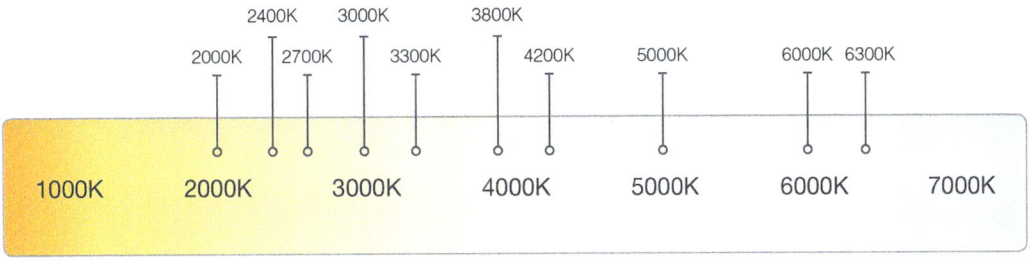

▲ Figure 2.8

Correlated Color Temperature. Chart shows Correlated Color Temperature (CCT) in degrees Kelvin (K). Courtesy of Elemental LED.

Kelvin or K scale (rather than describing the more complex spectral power distribution). However, while CCT is widely used to classify lighting, especially LEDs, as a metric it is simplistic compared to the spectral power distribution.[16] CCT ranges are considered to be visually warm (<3000 K), neutral (3500–4000 K), and cool (>4000 K).[17] (See Figure 2.8.)

Research has suggested significant benefits in object detection when lit by white light (containing a broader spectral distribution). LEDs with a CCT of 4,100 Kelvin provided the longest object detection distance, significantly greater than HPS luminaires of higher wattage.[18] LED lighting sources have a much broader color profile, with the additional color contrast helping increase detection distances. However, some pedestrians have a more positive reaction to lower, warmer CCT. Also, some research suggests that driver vision may be negatively affected by the adaptation of the eye when moving from darkness into an area lit by sources with heavy white or blue content.[19]

2.2.6 Street Environment and Street Trees

The need for and effectiveness of street lighting is substantially affected by the street environment. Street environments differ in ambient lighting and by components of the street right-of-way, such as trees, street furniture, and median islands.

Although trees have major benefits for the natural environment and as amenities for pedestrians, the potential of the tree canopy to interfere with desirable light levels needs to be considered. A Minnesota study found that summertime foliage on deciduous trees reduced horizontal light levels from street lights 19–33 percent and vertical levels 21–65 percent by comparison to winter conditions with leafless trees.[20]

2.2.7 Vehicle Headlights

Vehicle headlights are an important light source on the street. While headlamps can furnish sufficient light for driving in some cases, they also can be a source of glare. Motor vehicle headlamps alone (without street lighting) typically provide vertical illuminance of one lux (substantially lower than recommended for crosswalks) for an extended distance in the front of the vehicle. Headlights generally allow visibility for a distance greater than the required **stopping sight distance** for motor vehicles on straight sections of local residential streets traveling under 50 kilometers per hour (30 miles per hour).[21]

Vehicle headlight design is an important focus for safety advocates. The National Transportation Safety Board in a broad review of pedestrian safety issues stated: "The most feasible approach to improving lighting is to improve headlights on cars so drivers can better see and avoid pedestrians."[22] A more detailed discussion of recent and potential advances in headlight design is included in Chapters 5 and 10.

2.3 Key Reference Documents

The following represent key technical reference documents for this topic, focusing on street lighting. Other resources, such as placemaking, are discussed in later chapters. Standards, guidelines, and research summaries are frequently updated. Thus, some of these documents may be revised soon after publication of this *Guide*.

2.3.1 IES RP-8-18. *Recommended Practice for Design and Maintenance of Roadway and Parking Facility Lighting*[23]

This 2018 *Recommended Practice* provides an extremely detailed and comprehensive treatment of lighting design techniques and criteria for streets, highways, and parking facilities. It was prepared by the Illuminating Engineering Society (IES), the leading professional organization for North American lighting engineers and other specialists, and approved through an American National Standards Institute (ANSI) process. The document is a compilation of several IES standards and other documents of leading professional organizations. Part 1 on Fundamentals addresses:

- Lighting theory
- Calculations
- Obtrusive light
- Design process
- Standards and codes
- Use of computer software
- Maintenance and operations

Part 2 on Design addresses design considerations, issues, recommendations, criteria, and calculations. It provides numerous design examples and citations.

IES standards and a range of publications are now included in the IES Lighting Library.[24] This allows automated updating of technical references for the users, as well as cross-referencing, portability, bookmarking, and reader notes.

2.3.2 Transportation Association of Canada Guide for the Design of Roadway Lighting[25]

The Transportation Agency of Canada (TAC) is an organization with corporate and agency members focused on road infrastructure and urban transportation. Its 2006 *Roadway Lighting Guide* which includes some off-roadway facilities, is also divided into two volumes: fundamentals and design. The fundamentals volume covers:

- Vision and fundamental concepts
- Obtrusive lighting
- Planning and design process
- System components and common design elements
- Standards and codes
- Computer applications
- Maintenance

The second volume on design includes warrants determining the need for the lighting of arterial, collector, and local streets; underpasses, overpasses, and bridges; roundabouts; and midblock crosswalks. (These are discussed in more detail in Chapter 7.) Similar to IES RP 8-18, the TAC Guide includes lighting design criteria, with recommended light levels for sidewalks, other pedestrian/ bicycle paths, intersections, roundabouts, and midblock crosswalks. Finally, the TAC Guide provides a number of design examples.

2.3.3 AASHTO Lighting Design Guide[26]

The American Association of State Highway and Transportation Officials (AASHTO) is the primary organization for state transportation and highway departments. It issues an influential guide for the geometric design of roadways and streets. The companion *Lighting Design* volume was reissued in 2018.

The AASHTO Guide addresses warranting conditions, primarily for highways and bridges. It also addresses "Master Lighting Plans" in detail. In this context, Master Lighting Plans are addressed primarily as technical systems plans for an area or corridor lighting, concerning topics such as light levels, curfews, and dimming. Other topics in the *Guide* include pole placement, roundabouts, maintenance, and handling objectionable light.

2.3.4 FHWA Roadway Lighting Handbook[27]

The Federal Highways Administration (FHWA) *Roadway Lighting Handbook* provides guidance to designers and state and local agency staff on the applications of roadway lighting. This *Handbook* has no official standing as a standard or requirement, although it discusses how the FHWA considers lighting in administering funding. The document is intended as a supplement to AASHTO and IES documents. It covers:

- Policy and guidance
- Basic terms and concepts
- Warranting criteria (including the TAC warrants for streets and intersections)
- Lighting impacts
- Lighting design and equipment selection
- Application considerations;
- Other roadway systems (roadway markings and vehicle headlamps)

2.3.5 CIE Guidelines and Standards

The International Commission on Illumination (CIE in its French acronym) furthers global cooperation and information exchange on lighting, color vision, and image technology. CIE Division 4 focuses on lighting for transportation, producing international standards on roadway lighting, signs, signals, and light trespass. U.S. designers generally give AASHTO, IES, and FHWA guidance precedence over CIE standards. However, CIE research on innovative topics such as adaptive or smart lighting, LEDs, and measurement systems is of special interest. For example, CIE 236 summarizes empirical data regarding lighting for pedestrian safety, security, and reassurance.[28]

2.3.6 FHWA Midblock Crosswalk Lighting Informational Report[29]

The Virginia Tech Transportation Institute (VTTI) produced the FHWA *Midblock Crosswalk Lighting Informational Report*. This document provides information on lighting parameters and design criteria for the installation of lighting for midblock crosswalks. This covers vertical illumination, luminaire selection and placement, crosswalk placement, glare, and ambient lighting. It is based partly on static and dynamic tests of driver performance for detecting pedestrians and objects in midblock crosswalks. It also briefly considers lighting for intersection crosswalks.

2.3.7 IES LP-2-20. Lighting Practice: Designing Quality Lighting for People in Outdoor Environments[30]

A recent IES document provides guidance for lighting pedestrian spaces, particularly those outside the street right-of-way, such as paths through public parks and plazas. LP-2-20 is a Lighting Practice document, approved by an American

National Standards Institute (ANSI) process. It describes principles of effective lighting, which should satisfy a pyramid of hierarchical needs, starting with considerations of lighting zone and environmental context at the base, and then including orientation and wayfinding, reassurance, safety, atmosphere, and enjoyment. The approach also aims to minimize adverse impacts of lighting on the environment.

LP-2-20 describes considerations for lighting design (such as glare and color of light), as well as steps in the design process. As of publication, IES was actively considering issuing additional detailed guidance on how to implement LP-2-20, which itself does not provide detailed technical guidelines, such as light level recommendations for different types of facilities.

2.3.8 National Academies' Solid-State Roadway Lighting Design Guide[31]

This report provides guidance for the use of solid-state (primarily LED) lighting on roadways. It was developed through the National Cooperative Highway Research Program (NCHRP), an arm of the Transportation Research Board, which is government-funded but independent. The report is aimed mainly at engineers in state and local agencies involved in the design, management, and maintenance of roadway lighting. It addresses the special characteristics, advantages, and disadvantages of LED lighting, compared to the High-Intensity Discharge (HID) lighting that LEDs are rapidly replacing.

Volume 1 is a proposed AASHTO guide to supplement the AASHTO *Roadway Lighting Design Guide*, 7th edition. Volume 2 provides background data and analysis. This work includes a comprehensive literature review; a survey of practices of state and local transportation agencies regarding design, construction, and maintenance of LED lighting; and field research at the Virginia Tech Transportation Institute's test track.

2.4 Legal Responsibility

Courts in the majority of U.S. states take the position that unless there is a specific hazard known to the municipality, the city has no broad legal duty to provide lighting on city streets.[32] For example, the California Court of Appeals agreed with this position in a 1999 case.[33] Also, an Illinois case indicated in that state there is no broad duty to install traffic signs or street lights.[34]

Liability varies by state. For example, in Maine, courts have ruled that cities are not liable for injuries connected with insufficient lighting on sidewalks near municipal buildings.[35]

The Rhode Island Trust, a risk management association, advised its municipal members "to undertake traffic engineering reviews and address known hazards in areas where street lighting would be decreased" (typically to reduce electricity use). A Virginia case ruling held that if a city negligently **created** a nuisance not specifically authorized by law, the city can be found liable.[36] In this case, the city constructed a cul-de-sac near a creek, but did not install street lighting nor warning or protective devices to prevent vehicles from running off the road.

Cities typically do not have a broad legal duty to provide street lighting, but they can be found negligent for failure to maintain lighting properly or for creating a hazard to which a lack of lighting contributes. However, legal responsibilities vary by state.

Generally, once lights are installed, the city is liable if the lights are not maintained. Maintenance activities themselves are typically not protected to the same extent as improvements made following an appropriate design process.

Liability for municipalities and businesses from new "smart city" technologies (smart lighting including sensors on light poles, discussed generally in Chapter 6) involves complicated issues, such as privacy protections.[37] Promises of enhanced safety through innovative devices could increase municipal liability, but this caution alone should not deter municipalities from using new technology.[38] Documenting the basis for decisions on innovative lighting measures can be critical in defending against litigation.[39] This should include: providing research data on the effectiveness and impacts of measures; demonstrating adherence to standards and guidelines; and documenting design reviews by experts, such as licensed engineers.

Notes

1. Illuminating Engineering Society (IES), *RP 8-18: Recommended Practice for Design and Maintenance of Roadway and Parking Facilities Lighting* (New York: IES, 2018).
2. Peter Boyce, *Human Factors in Lighting*, 3rd edition (Boca Raton, FL: CRC Press, 2014), 497–502; Gunilla Haegerstrom-Portnoy, Marilyn E. Schneck, and John A. Brabyn, "Seeing into Old Age: Vision Function beyond Acuity," *Optometry and Vision Science* 76, no. 3 (1999): 141–158.
3. Clanton & Associates for City of San Jose, *Public Streetlight Design Guide* (San Jose: City of San Jose, 2016).
4. Sima Tawakoli, "Shadows and Light," IES Street and Area Lighting Conference, Dallas, Texas, October 28, 2020.
5. Steve Fotios et al., *CIE 236: Lighting for Pedestrians: A Summary of Empirical Data* (Vienna, Austria: International Commission on Illumination, 2019), doi: 10.25039/TR.236.2019.
6. IES, *Recommended Practice for Lighting Roadway and Parking Facilities*, 3–17.
7. Mark Rea, "Practical Implications of a New Visual Performance Model," *Lighting Research and Technology* 18, no. 3 (1986): 113–118; John D. Bullough, "New Approaches to Lighting for Pedestrian Safety and Sense of Personal Security," Transportation Research Board Human Factors Workshop on "Walking at Night: The Pedestrian's Perspective," January 8, 2017, http://www.pedbikeinfo.org/trbped/documents/2017/Bullough-TRBWorkshop-08Jan2017.pdf.
8. Paul Mitchell, "BUG Ratings," IES Street and Area Lighting Conference, Dallas, Texas, October 28, 2020.
9. Fotios et al., *CIE 236*.

10. "Photometric Lighting Plan," LED Lighting Supply Website, accessed April 15, 2021, https://www.ledlightingsupply.com/photometric-plan.

11. See https://www.ies.org/lighting-library/.

12. Steve Fotios et al., *CIE 236*.

13. Illuminating Engineering Society, *Lighting Practice LP-2-20: Quality Lighting Design for People in Outdoor Environments: An American National Standard* (New York: IES, 2021).

14. Illuminating Engineering Society, *Guideline (G) 1-16: Security Lighting for People, Property and Critical Infrastructure* (New York: IES, 2016).

15. Illuminating Engineering Society, *Technical Memorandum 30-18: IES Method for Evaluating Light Source Color Rendition* (New York: IES, 2018).

16. National Academies of Sciences, Engineering, and Medicine, *Solid State Roadway Lighting Design Guide: Volume 1: Guidance* (Washington, DC: The National Academies Press, 2020), 3, https://www.nap.edu/catalog/25678/solid-state-roadway-lighting-design-guide-volume-1-guidance.

17. Gary Meshberg, "Demystifying Tunable White," *Lighting Design + Application* 50, no. 2 (February 2020): 10.

18. IES RP-8-18, 2–9; Ronald Gibbons, "Connected Infrastructure Activities," Webinar on "Lighting for Pedestrian Safety and Walkability," October 17, 2018, http://www.pedbikeinfo.org/webinars/webinar_details.cfm?id=13.

19. National Academies, Solid-State Lighting, 4.

20. Patrick Hasson et al., "Trees, Lighting, and Safety in Context-Sensitive Solutions," *Transportation Research Record* 2120, no. 1 (January 2009): 101–111. doi: 10.3141/2120-11.

21. IES, *Recommended Practice for Design and Maintenance of Roadway and Parking Facility Lighting*, 3–6.

22. National Transportation Safety Board, *Special Investigation Report: Pedestrian Safety* (Washington, DC: NTSB, 2018), 18.

23. IES, *Recommended Practice for Design and Maintenance of Roadway and Parking Facility Lighting*.

24. "The Lighting Library® – Illuminating Engineering Society." https://www.ies.org/lighting-library/.

25. Transportation Association of Canada, *Guide for the Design of Roadway Lighting* (Ottawa, Canada: TAC, 2006).

26. American Association of State Highway and Transportation Officials, *Roadway Lighting Design Guide*, 7th Edition (Washington, DC: AASHTO, 2018).

27. Paul Lutkevich, Don McLean, Joseph Cheung, *FHWA Roadway Lighting Handbook*, FHWA SA-11-22 (Washington, DC: FHWA, 2012), https://safety.fhwa.dot.gov/roadway_dept/night_visib/lighting_handbook/.

28. Steve Fotios et al., *CIE 236*.

29. Ronald Gibbons et al., *Informational Report on Lighting Design for Midblock Crosswalks* (McLean, VA: FHWA Office of Safety Research & Development, 2008).

30. IES, *Lighting Practice: Outdoor Environments*.

31. National Academies, *Solid State Lighting*.

32. Rhode Island Interlocal Risk Management Trust, "Considering Reducing Streetlighting to Save Money?" *Risk Advisor*, September 2011, http://www.ritrust.com/Content/Files/public/Risk%20Advisor%20Streetlight%20Liability%20PDF.pdf.

33. "Plattner v. City of Riverside," *FindLaw*, https://caselaw.findlaw.com/ca-court-of-appeal/1129807.html.

34. Parsons v. Carbondale Township, 577 N.E. 2d 779 (Ill. App. 1991), https://www.mwl-law.com/wp-content/uploads/2013/03/MUNICIPAL-COUNTY-LOCAL-GOVERNMENTAL-LIABILITY-CHART-00212510.pdf.
35. Edward R. Benjamin, "Liability for Sidewalk Defects," *Maine Townsman*, Maine Municipal Association, January 2015.
36. Virginia Transportation Research Council, *Tort Liability: A Handbook for Employees of the Virginia Department of Transportation and Virginia Municipal Corporations*, Report 04-R30 (Charlottesville, VA: VTRC, 2004), http://www.virginiadot.org/vtrc/main/online_reports/pdf/04-r30.pdf.
37. Mark S. Raffman, "'Smart Cities' Raise Novel Issues and Novel Risks," *Yahoo Finance*, October 23, 2018, https://finance.yahoo.com/news/apos-smart-cities-apos-raise-050033962.html.
38. Doug Fraser, "Shark Detection System Raises Liability Concern in Massachusetts," *Cape Cod Times*, April 8, 2019, https://www.govtech.com/public-safety/Shark-Detection-System-Raises-Liability-Concern-in-Massachusetts.html.
39. Gibbons, Cheung, and Lutkevitch, *The Future of Roadway Lighting*.

Chapter 3

Benefits of Improving Lighting

3.1 Purpose and Scope of This Chapter

This chapter presents potential benefits of outdoor lighting improvements: safety, security and comfort, sense of place, aesthetic satisfaction, information, and increased economic activity. This broad range of benefits should be considered when prioritizing lighting versus some measures that only have pedestrian safety benefits. Many of the potential benefits overlap or are closely related. For example, providing an enhanced sense of security or aesthetic satisfaction can lead to more walking (with indirect physical and mental health benefits) and increased economic activity.

This chapter summarizes quantitative information on the range of benefits that are available. There is generally more information available on traffic safety than other potential benefits. But even for traffic safety, there is limited information associating benefits with specific types of lighting or lighting strategies. Traffic safety depends primarily on the pedestrian being visible to motorists to avoid collisions, while safety from tripping and slipping primarily depends on the visibility of objects to the pedestrian.

Chapter 4 next presents the financial costs and potential adverse impacts of lighting so the balance can be considered when planning and designing lighting enhancements. Some lighting improvements could also be considered to have indirect benefits by reducing adverse impacts of existing lighting, such as converting street lighting to light-emitting diodes (LEDs) to reduce electricity use. Additional information on benefits and potential adverse impacts is also presented on specific technologies in Chapters 5 and 6. Methods to increase benefits and limit adverse impacts are next addressed through planning and policy activities (Chapter 7) and transportation design, operations, and maintenance activities (Chapter 8). Chapter 9 also discusses more extensively how to derive placemaking and aesthetic benefits for outdoor lighting, with several case studies.

3.2 Duration of Darkness

In considering the potential benefits of lighting, it is important to note that low natural light conditions often extend throughout most of the 24-hour day cycle, depending on latitude and season. For example, in Seattle in mid-December,

DOI: 10.4324/9781003149750-3

full daylight hours are only roughly between 7:50 AM and 4:20 PM, so that much of the morning and evening commute peak periods are in twilight conditions or even full darkness.[1] Children commonly walk to or from schools outside daylight hours.

3.3 Traffic Safety

3.3.1 Fatalities and Injuries after Dark

Pedestrian fatalities and injuries are disproportionately higher after dark. Research using 11 years of U.S. Department of Transportation Fatality Analysis Reporting Systems (FARS) data on the distribution of fatal crashes estimated the safety risk to pedestrians to be at least four times higher in darkness than in daylight.[2]

Pedestrian collisions also tend to be higher in the fall and winter months. (See Figure 3.1 for Seattle example.) While there are multiple factors, including weather, shorter daytime hours are likely a factor. The change from Daylight

The risk of injury and especially fatality to pedestrians is much greater after dark. Low light conditions often extend through-out most of the 24-hour cycle, especially in winter and in northern latitudes.

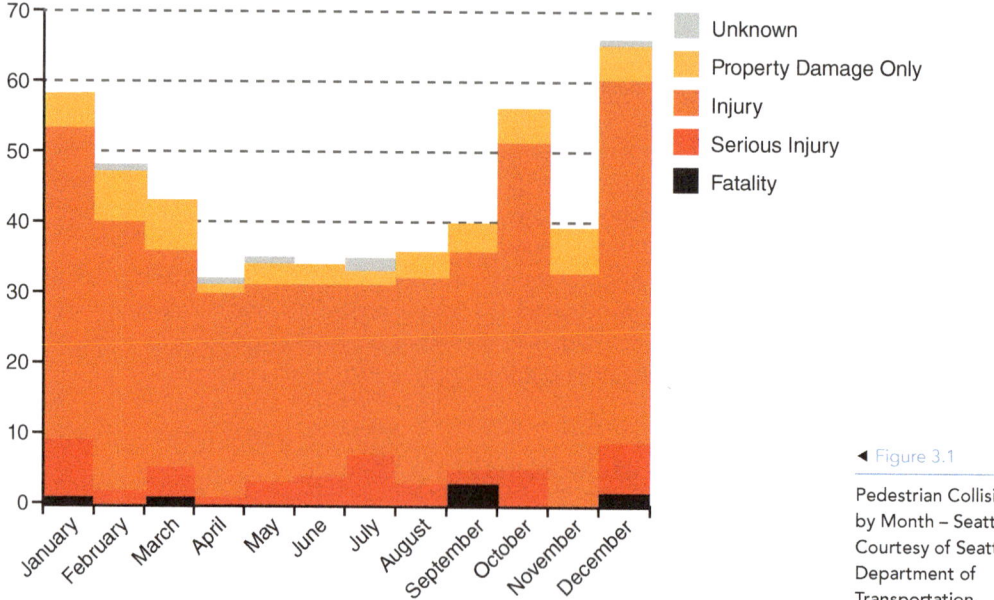

2015 PEDESTRIAN COLLISION SEVERITY BY MONTH

Unknown
Property Damage Only
Injury
Serious Injury
Fatality

◀ Figure 3.1

Pedestrian Collisions by Month – Seattle. Courtesy of Seattle Department of Transportation.

Savings Time to Standard Time is generally associated with a spike in pedestrian injuries. In fact, based on multivariate analyses of county-level data from the Fatality Analysis Reporting System for 1998 and 1999, researchers forecast that full-year daylight saving time would reduce pedestrian fatalities by 13 percent in the 5:00–10.00 AM and the 4:00–9:00 PM time periods.[3]

Pedestrian fatalities on U.S. roads increased by 53 percent over the 2009–2018 decade, while deaths of people riding in cars, trucks, and other motor vehicles remained essentially unchanged.[4] Pedestrians as a proportion of all traffic fatalities increased from 12 to 17 percent, the largest proportional increase among 30 developed nations. Nighttime pedestrian deaths accounted for 87 percent of the increase from 2009 to 2018. (Pedestrian deaths in urban areas, on arterials, and at non-intersection locations without crosswalks all had substantial increases.) While pedestrian fatalities decreased slightly from 2018 to 2019, pedestrian deaths as a percentage of all traffic fatalities remained constant.[5]

Experts suggest the largest factors in the increase are driver and pedestrian distraction (often due to devices like smartphones) and larger vehicles (especially SUVs and pick-up trucks).[6] As motor vehicles have become less vulnerable and safer for occupants (through size and measures such as improved airbags, brakes, and collision avoidance features), pedestrians have not benefitted as much from these changes. In fact, increased feelings of security by drivers may lead to reduced attention to pedestrians and the external environment, which could be especially hazardous after dark. Alcohol impairment was reported in about half of pedestrian fatalities.[7] About a third of pedestrians 16 years or older who died had a blood alcohol level of 0.08 or higher, and an estimated 13 percent of drivers in such crashes had blood alcohol at that level. Impaired driving and walking likely play a part in the disproportionate nighttime fatalities, but it is not clear how this relates to the increase. (Wide streets with poor crossing infrastructure and insufficient enforcement are also undoubtedly important factors, but it is unlikely these have changed proportionally to the increase in pedestrian fatalities in recent years.) Unfortunately, reliable data are very limited on many of these factors, such as driver/pedestrian distraction.

3.3.2 Reduced Fatalities and Injuries through Enhanced Lighting

The Vision Zero Network, a national group advocating a goal of eliminating traffic fatalities, has suggested that "clearly, improved lighting should not be overlooked as a promising strategy for Vision Zero communities. In fact, this may be one of the more overlooked areas to boost the movement toward safety for all."[8]

Several older studies have found that nighttime pedestrian injuries are typically reduced by roughly half by **adding illumination** to previously unlit roadway locations.[9] However, these meta-analyses combine studies of varying quality. There is also "test track" research indicating improved driver detection of pedestrians or targets with overhead roadway lighting. While vehicle headlights are generally sufficient to illuminate pedestrians at distances greater than

The addition of illumination to intersections significantly reduces vehicle/pedestrian collisions. There is some research available identifying the most important lighting factors for pedestrian safety.

stopping sight distance for speeds up to about 50 kilometers per hour (30 MPH), roadway lighting can provide greater peripheral visibility for pedestrians entering the roadway and also counter the glare from the headlights of oncoming vehicles.

The most systematic collection of **data assessing highway safety treatments** (or "countermeasures") is the Crash Modification Factor (CMF) Clearinghouse, maintained by the University of North Carolina's Highway Safety Research Center, with funding from the Federal Highway Administration. (Results are available online at www.CMFClearinghouse.org.) Four "high quality" research studies identified by the Clearinghouse found that the addition of illumination to previously unlit intersections resulted in at least a 42 percent reduction of nighttime vehicle/pedestrian collisions.

Detroit undertook a major LED installation project with 65,000 street lights, responding in part to conditions in which in 2014 roughly 40 percent of the street lights were not working. The Detroit Greenways Coalition stated that "public lighting improvements appear to be the primary factor behind Detroit's dropping [pedestrian] fatality rate."[10] Annual pedestrian fatalities declined from an average of 45.7 in the 2013–2015 period to 32.0 in the 2016–2019 period as lighting was reinstalled.[11]

Virginia Tech Transportation Institute (VTTI) researchers assessed various **factors affecting the visibility of pedestrians.** They found that, when compared to headlamps alone, roadway lighting doubles the distance at which a pedestrian is visible on a test track.[12] In a later VTTI study, the chance of a test track driver detecting a pedestrian at a safe distance was 1.7 times greater with a "high" level of surface luminance (1.5 candelas/square meter) compared to a "low" level (0.7 candelas/square meter).[13] Higher shoulder illuminance (80 percent of lane illuminance, or surround ratio of 0.8) resulted in a 2.6 times greater likelihood of safe detection than low shoulder illuminance (45 percent of lane illuminance). Lower uniformity led to a 2.3 times greater likelihood of safe detection. (Lower uniformity was a ratio of average luminance to minimum luminance of 1.8–3.5, while higher uniformity was a ratio of 1.3–1.4, which is more uniform than typically recommended levels.)

However, some **non-illumination factors** had even greater impacts than lighting. Younger drivers were 5.6 times more likely to detect the pedestrian at a safe distance. Drivers at 35 MPH (56 kilometers per hour) were 37 times more likely to detect pedestrians safely than drivers at 55 MPH (88 KPH). (Lighter-colored clothing also had a statistically significant but limited impact on pedestrian detection.)

Increasing **illumination levels** beyond a certain point does not improve safety substantially. VTTI researchers analyzed the safety impacts of roadway lighting as background for developing design criteria for adaptive roadway lighting.[14] They used night-to-day crash rate as the primary metric for safety impacts, calculating the relationship between horizontal illuminance levels and night-to-day crash rates, based on analysis of more than 88,000 crashes from 2004 to 2010. While the crash rate ratio dropped as horizontal illuminance increased from 0 to roughly 5–15 lux, increasing illuminance greater than 15 lux appeared to have minimal if any impact on crashes. (See Figure 3.2.) Horizontal illuminance is typically used in criteria for lighting intersections outside crosswalks.

▶ Figure 3.2

Relationship between Horizontal Illuminance and Night-Day Crash Rate Ratio. Courtesy of Virginia Tech Transportation Institute.

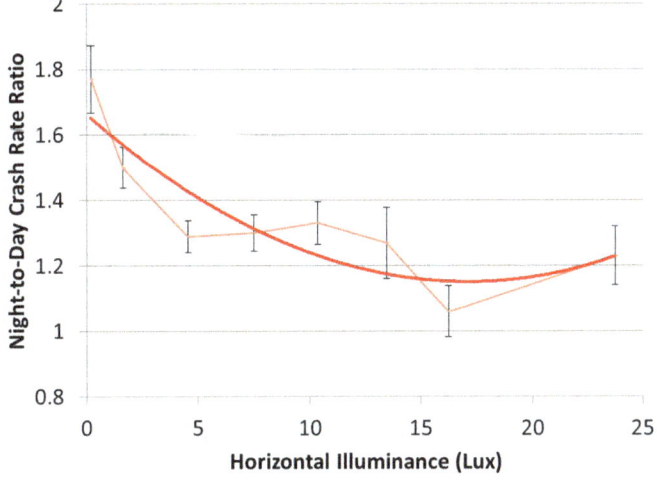

▶ Figure 3.3

Relationship between Vertical Illuminance and Night-Day Crash Rate Ratio. Courtesy of Virginia Tech Transportation Institute.

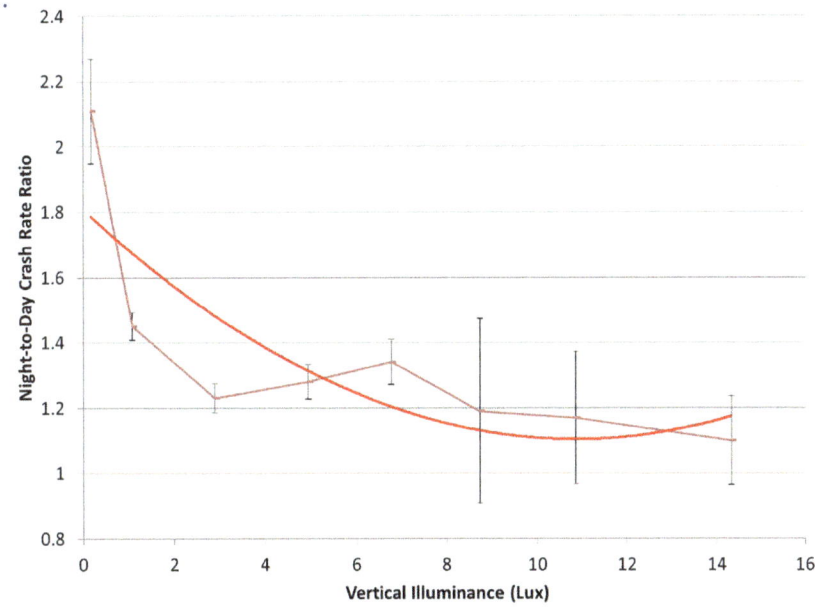

A similar analysis was performed for vertical illuminance, typically used in criteria for crosswalk illumination. This found increasing vertical illuminance from 0 to 4–10 lux was associated with a lower crash ratio, but further illuminance beyond 10 lux seemed to have little or no safety benefit. (See Figure 3.3.)

There is other research suggesting that **reducing street lighting** does not necessarily result in increased traffic collisions. British researchers found that none of the four strategies to reduce electricity consumption were associated with an increase in traffic collisions in 62 local jurisdictions over 14 years.[15] These strategies included switching selected street lights off completely, part-night lighting, dimming, and LED conversion. (Lighting changes were carefully planned by municipalities, and since 84 percent of jurisdictions analyzed had LED conversions, visibility in early evening may actually have been improved in some cases.)

3.3.3 How Lighting Improvements Complement Other Traffic Safety Measures

Poor or no lighting is rarely cited as the single **primary collision factor** in a pedestrian-involved crash, but it is often considered a contributing factor. (Drivers and pedestrians are implicitly expected to modify their behavior according to lighting and weather conditions.) For example, a Vision Zero San Francisco report on 2018 traffic fatalities found that the most commonly cited "primary collision factor" by the San Francisco Police Department was drivers' failure to yield to pedestrians at crosswalks.[16] (Of course, drivers may have failed to yield to pedestrians they did not see due to insufficient illumination.) The next most common factors were driver speeding and red light running. The San Francisco PedSafe study found a disproportionate share of pedestrian injuries caused by left-turning drivers failing to yield to pedestrians at signalized intersections without left turn phases.[17]

Low lighting intensifies the hazards of **speeding** motor vehicles for pedestrians. Researchers found that the posted speed limit was the dominant predictor of the dark/light crash ratio for pedestrian fatalities on U.S. highways.[18] The dark/light crash ratio was higher for higher-speed movements (straight roadway sections compared to curves, through movements compared to turning movements, and mid-block sections compared to intersections).

Vehicle speed is critical for determining whether a pedestrian survives a crash. If hit at 23 MPH (37 KPH), 90 percent of pedestrians survive, whereas among those hit at 50 MPH (80 KPH), only 20 percent survive.[19] The likelihood of completely avoiding an impending crash is naturally also higher at lower speeds, and pedestrians are more comfortable crossing or walking near lower-speed traffic.

Lighting improvements for pedestrians need to be prioritized along with **other potential safety measures**. For example, researcher Robert Schneider suggested that lower nighttime speed limits and increased after-dark enforcement should be considered to help address the large increase in nighttime pedestrian fatalities in the U.S. in the last decade, along with broader measures.[20]

The 2012 *FHWA Lighting Handbook* indicated that lighting improvements are a "proven safety countermeasure."[21] However, when the FHWA designated "proven countermeasures" in 2017, the list did not include street lighting.[22] Those specifically focused on **pedestrians** in the list included:

- Walkways
- Pedestrian hybrid beacons
- Pedestrian refuge islands and medians
- Leading Pedestrian Intervals (LPI or "ped head starts")

Pedestrian hybrid beacons are traffic control devices that use several overhead lights with a "wig wag" flash pattern and solid red indications to alert drivers to stop for pedestrians in or approaching the crosswalk. There is limited additional illumination of pedestrians or the crosswalk. Pedestrian hybrid beacons have been found to reduce pedestrian-involved crashes 55 percent, increasing driver yielding close to that of a traffic signal.[23] Rectangular rapid-flashing beacons are a similar, but less expensive, device with stutter-flash lights next to a crossing warning sign to alert drivers to yield or stop. They have also been shown to increase driver yielding significantly.[24]

Other "proven countermeasures" (which can influence pedestrian safety but are not strictly aimed at pedestrians) included:

Speed Control
- Roundabouts (and traffic circles)
- Road diets (reduction of through lanes, often with widened sidewalks and/or bicycle facilities)

Other Intersection Controls
- Signal backplates with retroreflective borders
- Left and right turn lanes at two-way STOP-controlled intersections
- Reduced left turn conflict intersections
- Other STOP-controlled low-cost measures
- Increased yellow change intervals

3.4 Security, Comfort, and Falls Prevention

Even walking only a block after dark, pedestrians need lighting sufficient for wayfinding and orientation, to avoid potential tripping hazards or obstacles, and for protection from crime.[25] "Low lighting" was one of the primary barriers Seattle residents cited as discouraging walking after dark.[26] Research bears that relationship out internationally.[27] In cities such as San Francisco, residents point to a lack of lighting also as a barrier to comfortable use of public transit.

Lighting researcher Steve Fotios points to the importance of "reassurance" to pedestrians.[28] Reassurance (confidence when walking alone) is an

> Lighting is needed by pedestrians after dark for wayfinding and orientation, to prevent falls, and to feel safe from crime.

> Effective lighting contributes to "reassurance" (confidence when walking alone).

important factor in the willingness to walk at night. Lighting can heighten reassurance directly by improving the visibility of landmarks, hazards, and criminal threats – and also indirectly by inducing more people to stay out after dark so they feel less alone. Pedestrians also may feel more reassured if they think the street scene is more visible to others, such as nearby residents.

3.4.1 Decreased Actual and Perceived Crime Rates

A comprehensive review of 13 studies concluded that improved street lighting significantly reduces crime.[29] The crime rate was reduced by an average of 38 percent in U.K. studies and 7 percent in U.S. studies. (Enhanced lighting, with even improved appearance of the equipment, was often associated with improved daytime personal security on affected blocks, perhaps by communicating to potential criminals that there is greater public attention to the location.) However, there was a broad range of results, with some studies finding no significant effect. More recently, a study of lighting reduction strategies in 62 localities in England and Wales found no significant association between "reduced" street lighting and crime, although it did not differentiate between daytime and nighttime crime.[30]

Key factors in perceptions of personal security for pedestrians at night include: a clear view of potential threats, with few if any hiding places, potential escape routes from danger, and the presence of others able to intervene or report problems.[31] When lab study participants were shown photo scenes and asked where they would be happy to walk alone, lighting was mentioned more frequently than clear views and places of refuge.[32] One factor in reassurance is the ability to recognize and evaluate other people on the street. A recent review found that six studies unanimously supported a significant relationship between illumination levels and facial recognition.[33]

3.4.2 Is Crime Just Being Relocated?

It is possible that increasing lighting on a particular block merely pushes criminals to adjacent blocks with lower lighting. A comprehensive review of street lighting and crime found that displacement usually did not occur or there was a net reduction in crime in the overall area.[34] Some criminals may also change the time of day for their infractions. Crime could in theory be pushed distant from the treatment area. Such areas are sometimes considered control or comparison areas by researchers, making conclusions challenging.

3.4.3 Visibility of Tripping and Slipping Hazards

Pedestrian falls are a major source of injury, especially for the elderly, more common than injuries from motor vehicles in Sweden and the UK.[35] Over 9,000 elderly pedestrian falls annually in the U.S. involve tripping on a curb.[36] In Victoria, Australia, a state of 6.4 million, an average of 1,680 hospitalizations and 3,545 emergency room visits annually were due to pedestrian falls in the street or road right-of-way.[37]

Visibility of obstacles is an essential factor in falls. University of Sheffield researchers conducted a series of experiments with raised blocks under different levels of illumination, with different light sources, and pedestrians of different ages. They found that increasing horizontal illuminance from 0.2 lux to 2.0 lux significantly improved detection, but further increasing illumination to 20 lux had much less of an effect.[38] At the lowest level of illumination, the likelihood of detection was double for younger pedestrians (under 45 years) under metal halide lights compared to older pedestrians (over 60 years) under high-pressure sodium lights. Color content of light (scotopic/photopic ratios) made a significant difference in detection at very low illuminance, but not at medium to high illuminance.[39] Researcher Peter Boyce (Rensselaer Polytechnic Institute) commented that a minimum illuminance in the range of 0.1–1.0 lux seems sufficient to protect 95 percent of pedestrians from tripping on a pavement misalignment of 25 mm.[40] This level of illumination is lower than IES recommendations for walkways even in low-density residential areas.

3.4.4 Determinants of Reassurance

An international research review found the perceptions of "reassurance" among nighttime pedestrians increased with higher levels of horizontal illuminance, but that above 10 lux the gains were very limited.[41] It also concluded that brighter and bluer-content lighting (with higher scotopic/photopic ratio, and higher scotopic lumens) enhanced feelings of security. Illumination needs to be sufficient to help pedestrians determine physical and facial features of a potential threat. Finally, the review suggested that more uniform illumination levels, with some lighting covering foliage adjacent to walkways, were most helpful. Research indicates that minimum illuminance and uniformity appear to be more reliable predictors of reassurance than average illuminance.[42]

Recent research suggests that uniformity contributes more to pedestrian reassurance than horizontal illuminance.[43] (The proportion of variation in the reassurance rating that could be predicted by the uniformity ratio was 85 percent versus 54 percent that could alternatively be predicted by horizontal illuminance.) Research in Australia also suggested that women generally felt safer in areas with greater light uniformity, warmer color appearance (lower CCT), higher color rendition, and higher luminance with lighting tailored to the pavement and spaces being lit.[44] The areas perceived as safer actually had lower mean illuminance values.

3.5 Sense of Place

Lighting is an important contributor to the sense of place for special districts, such as historic districts, tourist centers, major parks, and university campuses. Sense of place or uniqueness influences the attachment to a place and the feelings of belonging and having a clear mental map of a location. "Sense of place" benefits translate into economic advantages, such as increased visitor spending, as well as visitor and resident satisfaction and civic pride.

3.5.1 Historic Identity

Numerous tourist areas use ornamental luminaires and poles to contribute to a unique identity and increased vibrancy at night. For example, Vancouver's Gastown and San Diego's Gaslamp districts use historic streetlights to "brand" large areas. (See Figure 3.4.) In San Francisco, historic street lights on the two main commercial/ceremonial streets (Market Street and Van Ness Avenue) have

◄ Figure 3.4

Vancouver's Gastown District – Historic Lighting and Steam Clock. Courtesy of Adobe Stock Photos. © yooranpark - stock. adobe.com.

Lighting appears to be a significant economic contributor to tourist and commercial areas, but there has been minimal research on this.

been the focus of recent civic controversy over potential adverse impacts from street improvement projects on historic lighting equipment.

Historic or ornamental lighting is also popular in many older residential neighborhoods. For example, the City of San Jose City Council needed to provide guidance to staff on responding to numerous resident requests for new or replacement ornamental lighting. Some manufacturers specialize in historic-theme lighting equipment.

3.5.2 Place Relationship to Technology

Lighting can also be used to promote a "Tomorrowland" feeling of technological triumph. Programmable features and unusual equipment emphasize how lighting has advanced in recent years.

For example, Vancouver, British Columbia used programmed lighting and novel mountings and fixtures to provide an exciting sense of technological potential at a plaza designed for the 2010 Winter Olympics. The San Francisco-Oakland Bay Bridge Lights offer a programed light show visible to thousands of pedestrians and diners, especially those on the waterfront Embarcadero. (See Figure 3.5.) Designed by light artist Leo Villareal and sponsored by Illuminate SF, some 25,000 individually programmed white lights strung on bridge cables allow monumental displays of playful light cascades and dance moves. This calls attention to the relationship of this critical piece of infrastructure to the modern cities at either end, filled with high-tech employers.

▶ Figure 3.5

San Francisco Bay Bridge Lights. Courtesy of Adobe Stock Photos. © Sundry Photography - stock.adobe.com.

3.5.3 Place Relationship to the Natural Environment

Lighting can be adapted to blend in with natural surroundings. For instance, a park setting may be enhanced by lower levels of illumination and understated luminaires and mountings.

An example of such lighting was designed for the Grand Canyon visitor center by Clanton & Associates.[45] (See Figure 3.6.) To support popular stargazing activities, the park upgraded the lighting control system to dim walking path lighting systems to better accommodate the evening sessions. The installation complies with International Dark-Sky Association criteria. LED lights are controlled by touch screens, allowing park staff to adjust preset levels easily.

3.6 Aesthetics and Information

Light and shadow have been critical elements of the artist's palette for centuries for paintings and sculptures. Outdoor lighting makes it possible to appreciate the streetscape after dark. It facilitates viewing details of architecture, foliage, street furniture, and other sights. Outdoor lighting equipment and illumination can provide artistic touches to inspire appreciation and wonder. Light poles themselves may be attractive street furniture pieces, sometimes hosting flowers or banners. Lighting displays can also provide information.

A programmed light show on a bridge or high-rise building is visible for blocks, free to passersby, and less intimidating than a museum. The dynamic potentials of changing light levels, color appearance, and images are also advantageous. Street lighting provides nearly ubiquitous potential locations for broadcasting information. Potential disadvantages include light pollution (glare, sky glow, light trespass), cost, high maintenance needs, and difficulty in protecting installations from vandals. But these challenges can be mitigated through appropriate technical requirements, guidance, and design (e.g., vandal-proof materials).

◀ Figure 3.6

Grand Canyon Walking Path. Courtesy of National Park Service and Michael Quinn, staff photographer.

Lighting used for aesthetic and information purposes can provide large-scale, dramatic, and dynamic effects, such as highlighting important buildings or infrastructure. Illumination levels and lighting equipment can be designed to support moods or themes.

3.6.1 Public Art

"Light art" is a recognized form of modern artistic expression. Often installations are large-scale and dynamic (with changing colors, images, and textures). Artworks using the medium of light can communicate subtle, symbolic, or associative concepts. These artworks can attract pedestrians to a neighborhood or district, enhancing the sense of nighttime security and boosting patronage of restaurants, art galleries, and other businesses.

For example, Jim Campbell is an internationally recognized light artist, based in San Francisco, who works both on small museum pieces and large outdoor displays. His work has been recognized with shows at leading museums like the Metropolitan, Whitney, and San Francisco Museum of Modern Art. His best-known work is "Day for Night," in which the top nine stories of the tallest San Francisco building, the Salesforce Tower, house 11,136 LED bulbs.[46] They flash a dynamic mix of generic images like dancers and local scenes of clouds, bay life, waves, crowds, and birds. Campbell works with different levels of resolution, abstraction, and diffusion.

3.6.2 Information

Through images, simple verbal messages, and colors, lighting displays can provide information or visual messages to pedestrians about weather, special events, and the like. The large scale, dynamic nature, and simplicity of lighting communications can be an advantage after dark over static signs. For example, San Francisco's City Hall and the Empire State Building are often used as the backdrop for color projections related to groups or events. (See Figure 3.7.)

Street light poles can support public and commercial information provision through low-tech means such as banners and high-tech methods such as Bluetooth or cellular transmitters. Poles can illustrate historic episodes for the city or district (e.g., with mini-sculptures on their bases).

3.6.3 Fairs and Festivals

Artistic lighting displays are a key feature of many fairs and special events globally. Multiple light exhibits are often used to lead pedestrians through a district. Projections on buildings highlight the architecture of the city, with a monumental scale and dynamism. Such festivals generate significant economic benefits for

◀ Figure 3.7

Empire State
Building Colored
Lights Display.
Courtesy of Adobe
Stock Photos. ©
rabbit75_fot - stock.
adobe.com.

hosts and support international cultural exchanges. They can provide a venue for designers and researchers to test "smart lighting" features before they are used in permanent installations.[47]

Vivid Sydney (Australia) is one of the largest annual events. The 2019 festival included creative industry forums, free outdoor light sculptures and installations, and a cutting-edge music program. Images are projected on the iconic Sydney Opera House. (See Figure 3.8.) Light projections and sculptures are used for informational purposes, such as telling the story of wildlife conservation at the Taronga Zoo.

Other prominent light festivals include: the Berlin Festival of Lights, the Harbin (China) International Ice and Snow Sculpture Festival, the Festival of Lights Lyon (France), the Portland Winter Light Festival, the Amsterdam Light Festival, the Reykjavik Winter Lights Festival, and the St. Petersburg White Nights Festival.

Candles, lanterns, and festoon lights figure prominently in religious and civic holidays, reflecting the comfort and aesthetic pleasure they provide. Christmas light displays and Ramadan lanterns are ubiquitous in many countries.

The festival of Diwali is celebrated internationally with traditional lamps and modern light shows by more than a billion people, Hindus and other religious groups.[48]

3.7 Increased Economic Activity

There is anecdotal and some quantitative evidence that improved lighting can attract tourists and residents to particular districts. (The Broadway theater district's nickname "Great White Way," which dates back well over a century, suggests the power of lighting to draw nighttime patronage to an area.) Greater patronage supports spending on restaurants, theaters, and other establishments. Of course, the impacts of outdoor lighting are difficult to separate from numerous other factors, such as security, street cleaning, or business marketing efforts.

Paul Levy, the president and CEO of the Philadelphia Center City District stated that street lighting (with over 2,100 new light poles in a 152-block area installed) was one of the major factors in Downtown Philadelphia's resurgence in the period from roughly 1990 to 2010.[49] "More pedestrians ventured out, retailers and restaurants expanded evening hours, and bright windows and sidewalk seating further animated the street. A virtuous circle ensues as ever more people choose to live downtown near work and the thriving arts and entertainment scene." He pointed to the following increases from 1992 to 2002:

- Downtown restaurants open at night increased from 65 to 192
- Outdoor cafes increased from 0 to 104
- Hotel rooms increased 65 percent
- Over 4,000 apartments were created, with about 1,000 in the pipeline

Lyon, France, developed a citywide lighting plan in 1989 in part to showcase its historic buildings.[50] The plan included illuminating 300 sites, many of them historic. The Deputy Mayor of Lyon, Gilles Buna, stated that lighting guided by the plan is "a magnet for tourists, the pride of local citizens, an undisputed area of competence, and an economic lever." The plan was revised in 2005 to emphasize energy conservation and sustainable development, adding nearly 1,000 lights while reducing electricity use by a third.[51] District lighting plans have also been prepared and integrated into efforts to emphasize quality lighting in private development projects.

Lyon also sponsors an annual Festival of Lights, whose spiritual origin the City claims dates back to 1852, when residents placed candles in colored glass on their window sills to celebrate the installation of a religious statue.[52] The 2019 Lyon Festival of Lights attracted an estimated 1.8 million visitors over four evenings. "Vivid Sydney" also demonstrates the economic potential of such light festivals. The 2019 event attracted 2.4 million visits.[53] The 2018 event was credited with generating $173 million in tourism-related spending over a 23-day program.[54]

Notes

1. "Sunrise and Sunset Times in Seattle," Time and Date website, accessed March 6, 2021. https://www.timeanddate.com/sun/usa/seattle.
2. John M. Sullivan and Michael J. Flannagan, *Characteristics of Pedestrian Risk in Darkness*, Technical Report: UMTRI 2001–33 (Ann Arbor: University of Michigan Transportation Research Institute, 2001).
3. Douglas Coate and Sara Markowitz, "The Effects of Daylight and Daylight Saving Time on US Pedestrian Fatalities and Motor Vehicle Occupant Fatalities," *Accident Analysis & Prevention* 36, no. 3 (May 1, 2004): 351–357. doi: 10.1016/S0001-4575(03)00015-0.
4. B.C. Tefft, L.S. Arnold, and W.J. Horrey, *Examining the Increase in Pedestrian Fatalities in the United States, 2009–2018 (Research Brief)* (Washington, DC: AAA Foundation for Traffic Safety, 2021).
5. Richard Retting, *Pedestrian Traffic Fatalities by State: 2020 Preliminary Data* (Washington, DC: Governors Highway Safety Association, 2021), https://www.ghsa.org/resources/Pedestrians21.
6. Retting, *Pedestrian Fatalities*; Sea Stachura, "Why Pedestrian Deaths Are at a 30-Year High," National Public Radio website, March 28, 2019, https://www.npr.org/2019/03/28/706481382/why-pedestrian-deaths-are-at-a-30-year-high.
7. Retting, *Pedestrian Fatalities*, 20.
8. Kathleen Ferrier, "Webinar Recap: Lighting Can Provide for Safety AND Data in Vision Zero," Vision Zero Network website, April 27, 2018. https://visionzeronetwork.org/focus-on-lighting/.
9. R.N. Schwab et al., *Synthesis of Safety Research Related to Traffic Control and Roadway Elements*, Volume 2, Chapter 12: "Highway Lighting." Report No. FHWA-TS-82-233 (Washington, DC: Federal Highway Administration, 1982); R. Elvik, "Meta-Analysis of Evaluations of Public Lighting as Accident Countermeasure," *Transportation Research Record* 1485 (1995): 112–123; Commission Internationale de l'Éclairage (CIE), *Road Lighting as an Accident Countermeasure*, CIE No. 93 (Vienna, Austria: Commission Internationale de l'Éclairage, 1992).

10. Detroit Greenways Coalition, "Detroit Public Lighting Improvements Reducing Pedestrian Fatalities," July 16, 2018, https://detroitgreenways.org/detroit-public-lighting-improvements-reducing-pedestrian-fatalities/.

11. Updated data on pedestrian fatalities provided by Todd Scott, Detroit Greenways Coalition, Email, December 11, 2020.

12. National Academies of Sciences, Engineering, and Medicine, *Solid-State Roadway Lighting Design Guide: Volume 2: Research Overview* (Washington, DC: The National Academy Press, 2020), https://www.nap.edu/catalog/25679/solid-state-roadway-lighting-design-guide-volume-2-research-overview, 22.

13. National Academies Vol. 2, *Solid-State*, 60.

14. Ronald Gibbons, Joseph Cheung, and Paul Lutkevich. "The Future of Roadway Lighting," *Public Roads* 79, no. 3 (November/December 2015), https://www.fhwa.dot.gov/publications/publicroads/15novdec/06.cfm.

15. Rebecca Steinbach et al., "The Effect of Reduced Street Lighting on Road Casualties and Crime in England and Wales: Controlled Interrupted Time Series Analysis," *Journal of Epidemiology and Community Health* 69 (2015): 1118–1124, https://jech.bmj.com/content/69/11/1118.

16. San Francisco Department of Public Health, San Francisco Municipal Tranportation Agency, and San Francisco Police Department, *Vision Zero Traffic Fatalities: 2018 End of Year Report* (San Francisco: DPH, 2019), https://www.visionzerosf.org/wp-content/uploads/2019/02/Vision-Zero-2018-End-of-Year-Traffic-Fatalities_2.14.2019-1.pdf.

17. Frank Markowitz and David Ragland, "FHWA PedSafe: The San Francisco MTA/UC Berkeley Pedestrian Safety Program," FHWA Safety website, January 11, 2009, https://safety.fhwa.dot.gov/ped_bike/tools_solve/ped_scdproj/webinar052809/sf/.

18. J.M. Sullivan and M.J. Flanagan, "Determining the Potential Safety Benefit of Improved Lighting in Three Pedestrian Crash Scenarios," *Accident Analysis and Prevention* 39, no. 3 (May 2007): 638–647, https://www.sciencedirect.com/science/article/abs/pii/S0001457506001941?via%3Dihub.

19. Brian Tefft, *Impact Speed and a Pedestrian's Risk of Severe Injury or Death* (Washington, DC: AAA Foundation for Traffic Safety, 2011), https://aaafoundation.org/impact-speed-pedestrians-risk-severe-injury-death/#:~:text=The%20average%20risk%20of%20death,Risks%20vary%20significantly%20by%20age.

20. Robert J. Schneider, "United States Pedestrian Fatality Trends 1977 to 2016," *Transportation Research Record* 2674, no. 9 (September 2020): 1069–1083, https://journals.sagepub.com/doi/10.1177/0361198120933636.

21. Paul Lutkevich, Don McLean, and Joseph Cheung, *FHWA Lighting Handbook* (Washington, DC: FHWA Office of Safety, 2012), https://safety.fhwa.dot.gov/roadway_dept/night_visib/lighting_handbook/.

22. FHWA Office of Safety, "Proven Safety Countermeasures: Pedestrian Hybrid Beacons," FHWA-SA-17-065, 2017, https://safety.fhwa.dot.gov/provencountermeasures/ped_hybrid_beacon/.

23. FHWA Office of Safety, "Proven Safety Countermeasures."

24. Martin C. Knopp, Associate Administrator for Operations, FHWA, "Memo: MUTCD-Interim Approval for Optional Use of Pedestrian-Actuated Rectangular Rapid-Flashing Beacons at Uncontrolled Marked Crosswalks," March 20, 2018, https://mutcd.fhwa.dot.gov/resources/interim_approval/ia21/index.htm.

25. Steve Fotios, J. Unwin, and Steven Farrall, "Road Lighting and Pedestrian Reassurance after Dark: A Review," *Lighting Research & Technology* 47 (2015): 449–469, https://journals.sagepub.com/doi/10.1177/1477153514524587.

26. Seattle Department of Transportation, *Pedestrian Lighting Citywide Plan* (City of Seattle: SDOT, June 2012), http://www.seattle.gov/Assets/Documents/Departments/SDOT/About/DocumentLibrary/PedMasterPlan/PedLightingFINAL.pdf.

27. Sarah Foster et al., "Safe Residential Environments? A Longitudinal Analysis of the Influence of Crime-Related Safety on Walking," *International Journal of Behavioral Nutrition and Physical Activity* 13 (2016): 22–30; Phil Mason, Ade Kearns, and Mark Livingston, "Safe Going: The Influence of Crime Rates and Perceived Crime and Safety on Walking in Deprived Neighborhoods," *Social Science and Medicine* 91 (2013): 15–24.

28. Fotios, Unwin and Farrall, "Road Lighting and Pedestrian Reassurance."

29. Brandon C. Welsh and David P. Farrington, "Effects of Improved Street Lighting on Crime," *Campbell Systematic Reviews* 4, no. 1 (2008): 1–51, doi: 10.4073/csr.2008.13.

30. Steinbach et al., "The Effect of Reduced Street Lighting on Road Casualties and Crime in England and Wales."

31. Leon van Rijswijk and Antal Haans, "Illuminating for Safety: Investigating the Role of Lighting Appraisals on the Perception of Safety in the Urban Environment," *Environment and Behavior* 50, no. 8 (October 2018): 889–912, doi: 10.1177/0013916517718888.

32. Fotios, Unwin and Farrall, "Road Lighting and Pedestrian Reassurance."

33. Steve Fotios and Maria Johansson, "Appraising the Intention of Other People: Discussion of Ecological Validity and Procedures for Investigating Effects of Lighting for Pedestrians," *Lighting Research & Technology* 51, no. 1 (2019): 111–130.

34. Ronald V. Clarke, *Improving Street Lighting to Reduce Crime in Residential Areas* (Washington, DC: US Department of Justice Office of Community-Oriented Policing Services, 2008), 8, https://cops.usdoj.gov/RIC/Publications/cops-p156-pub.pdf.

35. CIE (International Commission on Illumination), *CIE 236: Technical Report: Lighting for Pedestrians: A Summary of Empirical Data* (Vienna, Austria: CIE, 2019).

36. Rebecca B. Naumann et al, "Older Adult Pedestrian Injuries in the U.S.: Causes and Contributing Circumstances," *International Journal of Injury Control and Safety Promotion* 18, no. 1 (2011): 65–73. https://www.tandfonline.com/doi/full/10.1080/17457300.2010.517321.

37. Jennifer Oxley et al., "Falling While Walking: A Hidden Contributor to Pedestrian Injury," *Accident Analysis and Prevention* 114 (May 2018): 77–82.

38. Steve Fotios and Chris Cheal, "Obstacle Detection: A Pilot Study Investigating the Effects of Lamp Type, Illuminance and Age," *Lighting Research and Technology* 41 (2009): 321, https://journals.sagepub.com/doi/full/10.1177/1477153515602954.

39. Jim Uttley, Steve Fotios, and Chris Cheal, "Effect of Illuminance and Spectrum on Peripheral Obstacle Detection by Pedestrians," *Lighting Research & Technology* 49, no. 2, (2017): 211–227.

40. Peter Robert Boyce, *Human Factors in Lighting*, 3rd ed. (Boca Raton, FL: CRC Press, 2014), 435.

41. Fotios, Unwin and Farrall, "Road Lighting and Pedestrian Reassurance."

42. Steve Fotios, Alexandra Liachenko Monteiro, and Jim Uttley, "Evaluation of Pedestrian Reassurance Gained by Higher Illuminances in Residential Streets Using the Day-Dark Approach," *Lighting Research and Technology* 51, no. 4 (2018): 557–575, doi: 10.1177/14477153518775464; Peter R. Boyce et al., "Perceptions of Safety at Night in Different Lighting Conditions," *Lighting Research and Technology* 32 (2000): 79–91.

43. Fotios, Liachenko, and Uttley, "Evaluation of Pedestrian Reassurance Gained by Higher Illuminances in Residential Streets Using the Day-Dark Approach."

44. Nicole Kalms and Tim Hunt, "More Lighting Alone Does Not Create Safer Cities," *The Conversation*, May 28, 2019, https://theconversation.com/more-lighting-alone-does-not-create-safer-cities-look-at-what-research-with-young-women-tells-us-113359.

45. "Grand Canyon Visitor Center," *Legrand Integrated Solutions* (blog), accessed March 6, 2021, https://www.legrandintegratedsolutions.com/case_history/grand-canyon.

46. Zahid Sardar, "Local Artist Jim Campbell Creates Dramatic LED Light Installations," *Spaces*, June 28, 2019, https://spacesmag.com/gallery/art/local-artist-jim-campbell-creates-dramatic-led-light-installations/.

47. M. Hank Haeusler, "The Sydney Vivid Festival: From Place Branding to Smart Cities," in Sandy Isenstadt, Margaret Maile Petty, and Dietrich Neumann, eds., *Cities of Light*, (New York: Routledge, 2015), 144–147.

48. N'dea Yancey-Bragg, "What Is Diwali, the Festival of Lights, and How Will It Be Celebrated Amid Coronavirus?" *USA Today*, November 11, 2020, https://www.usatoday.com/story/news/nation/2020/11/11/diwali-2020-what-festival-lights-and-how-celebrated/6235436002/.

49. Jamie Bratt et al., *Best Practices in Placemaking through Illumination* (Blacksburg, VA: Virginia Tech Urban Affairs and Planning Program, 2010), https://www.arlingtoneconomicdevelopment.com/index.cfm?LinkServID=8B1450DD-D628-416F-9C7FB8702740004E&showMeta=0.

50. Bratt et al., *Best Practices in Placemaking*, 29.

51. Thierry Marsick, "Lyon, France," in LUCI (Lighting Urban Community International), *Exploring City Nightscapes* (Lyon: LUCI, 2020).

52. Fete des Lumieres and Ville de Lyon, "The 2019 Festival of Lights: Facts and Figures," Fete des Lumiere website, accessed January 25, 2021, https://www.fetedeslumieres.lyon.fr/en/news/2019-festival-lights-facts-and-figures#:~:text=The%20City%20of%20Lyon%20joined,during%20the%20four%2Dday%20event.

53. "Vivid Sydney 2019 Breaks All Records for Visitation – Destination NSW," accessed March 6, 2021, https://www.destinationnsw.com.au/news-and-media/media-releases/vivid-sydney-2019-breaks-all-records-for-visitation.

54. "Five Days to Glow: Vivid Sydney 2019" (Vivid Sydney, May 20, 2019), https://www.vividsydney.com/sites/default/files/2019-05/Five-Days-To-Glow_Vivid-Sydney-2019.pdf.

Chapter 4

Costs and Potential Adverse Impacts of Lighting

4.1 Purpose and Scope of This Chapter

While there are clear benefits of illumination described in the previous chapter, this chapter presents potential challenges and downsides of lighting, starting with increased financial costs, energy use, and light pollution (light trespass, glare, and skyglow). It expands on the human health impacts and flora and fauna impacts of artificial light, as well as fixed object hazards, and adverse aesthetic impacts. Financial costs and other impacts are often directly related. For example, light-emitting diode (LED) lighting conversions reduce both energy use and costs.

Lighting specialists and design guidelines are generally quite sensitive to these issues. Fortunately, there is a substantial amount of information available on how different lighting types and strategies can minimize such adverse impacts. This information can be used to try to find the ideal of the "right dose" of lighting (in the words of lighting design expert Ronald Gibbons of the Virginia Tech Transportation Institute).[1] Methods of mitigating these impacts are discussed throughout the chapter. While public agencies and utilities have direct control over public lighting, the significant challenges of private sites and building lighting also need to be addressed. Tools for regulating private lighting are discussed in the final section.

Costs and other adverse impacts are also considered in future chapters. For example, Chapter 6 investigates new technologies, considering both benefits and potential negative impacts. Plan and policy development (Chapter 7) and transportation design/operations efforts (Chapter 8) generally aim to reduce costs and adverse impacts as a primary goal.

4.2 Financial Costs

For lighting, there are three major types of costs to be considered: the capital costs of installation, operations costs (primarily electricity use), and maintenance costs. Life cycle cost estimates consider all three categories. Some lighting costs are not readily apparent. For example, new light poles may trigger the need for additional tree trimming.

DOI: 10.4324/9781003149750-4

Lighting is more expensive than many other pedestrian safety measures, particularly when life cycle costs are considered. Lighting may also cause potential adverse impacts, such as light pollution. However, lighting can provide substantial health, social, and economic benefits that other safety measures generally do not.

4.2.1 Capital Costs

Capital costs include the design and installation of new or enhanced lighting. Lighting costs are generally similar to or higher than other pedestrian improvements for a block or intersection. For a spot location, lighting costs are much higher than signs and striping, but below or of similar magnitude as traffic signals, pedestrian hybrid beacons, and corner curb extensions. However, costs vary enormously depending on the extent and type of lighting, as well as local factors.

The initial capital cost of a particular lighting project depends on a complex array of factors, including:

1. The material cost of luminaires, poles, foundations, wiring, and the like
2. Labor costs for design, installation, and inspection
3. Site conditions, such as the proximity of an electrical power source (or sunlight for a solar project)
4. The cost of disposal of existing lights (if a replacement project)
5. Measures to avoid conflicts with existing or planned trees, street furniture, utilities, adjacent railroad lines, or airport restrictions
6. Measures or design criteria to minimize environmental impacts

Such factors require a detailed site and documents investigation as part of the design process in order to develop a precise cost estimate. The design is also subject to the lighting standards and aesthetic criteria determined by the local government or utility. Transportation planning projects and lighting master plans typically cannot investigate location-specific cost factors in detail, so unit costs from similar projects are more realistic for planning efforts.

Adding new lighting to a single crosswalk costs typically in the range of $11,000–$42,000.[2] A single pedestrian-scale pole and luminaire in the high-cost San Francisco Bay Area is estimated at $10,500–$13,000 for materials and installation labor.[3] On a per-mile basis, sample capital costs for pedestrian-scale lighting include:

- $1.8 million (San Francisco's Balboa Park pedestrian-scale lighting project in 2012, including design and other costs)[4]
- $3.3 million (Brookline MA, for mixed pedestrian-scale and roadway lighting 2020, including design and police support)[5]
- $704,000 (Alameda County, California, *2019 Bicycle and Pedestrian Master Plan* unit cost, only installation labor and materials, without design, assuming pedestrian-scale lighting spaced every 150 feet)[6]

Capital Costs of Typical Pedestrian Safety Countermeasures

Countermeasure	Unit	Mean Cost (in $2012 or earlier)	Maximum Cost (in $2012 or earlier)	Number of Sources	Number of Observations
Street Light	One Luminaire/Pole	$4,880	$13,900	12	17
In-Pavement Lights	One Crosswalk system	$17,620	$40,000	4	4
High Visibility Crosswalk	Each crosswalk	$2,540	$5,710	4	4
Curb Extension	Each bulb	$13,000	$41,170	19	28
Curb Ramp	Each	$810	$3,600	16	31
Rectangular Rapid Flashing Beacon	Each crosswalk	$22,250	$52,310	3	4
Pedestrian Hybrid Beacon	Each crosswalk	$57,680	$128,660	9	9
Median Refuge Island	Square foot	$10	$26	6	15
Pedestrian Signal (added to signalized crosswalk)	Each crosswalk	$1,480	$10,000	22	33
Roundabout/Traffic Circle	Each	$85,370	$523,080	11	14
Sidewalk	Square foot	$32	$410	46	164
Speed Hump	Each	$2,640	$6,860	14	14

Adapted from: Bushell et al. (2013).[8] Courtesy of University of North Carolina Highway Safety Research Center.

The University of North Carolina Highway Safety Research Center provided sample cost data for numerous bicycle, traffic calming, pedestrian measures, as well as general traffic engineering measures (signs, signals, striping). Example costs for a range of pedestrian measures are presented in Table 4.1. This report found a mean capital cost per street light of $4,880, with a maximum of $13,900, based on 12 sources.[7] The document did not distinguish between regular street lights and pedestrian-scale lights. This publication also reported in the "lighting" section a mean capital cost for in-pavement lights of $17,620, with a maximum of $40,000, based on four sources. This apparently referred to flashing in-pavement lights, a warning device to alert drivers, with only limited illumination of pedestrians.

Two other devices, the rectangular rapid flashing beacons (RRFBs) and the pedestrian hybrid beacon (PHB), similarly are considered traffic control devices that use primarily flashing lights to slow or stop drivers when pedestrians are in or approaching the crosswalk. They provide limited illumination of the pedestrians or crosswalks. Both use distinctive flashing patterns, like a stutter flash for the RRFB and a "wig wag" pattern for the PHB.

4.2.2 Operations and Maintenance Costs

Operations and maintenance costs for lighting are substantial, typically higher than other pedestrian improvements. Municipal street lighting is a major part of the typical city government operating budget. For example, the City of Los Angeles spends $42 million, about $188 per street light, on annual operations and maintenance.[9]

A principal operating cost component is the electricity supply. Street lighting can account for as much as 40 percent of a municipality's electric utility cost.[10] For example, the annual cost just for street light electricity was as high as $16 million for Los Angeles, a city of 4 million, but has been reduced by about $10 million by its recent LED conversion project.[11] Phoenix spent nearly $7 million for street light electricity before LED conversion for the city of 1.7 million, but this is expected to be reduced by $3.5 million to $2.5 million.[12]

Maintenance costs are also substantial. Street light components need occasional replacement. The variety of fixtures used throughout decades complicates maintenance; the City of Los Angeles, at the extreme, has over 400 styles. (And the City even supports a small street light museum.) Trees need to be trimmed to avoid blocking light. Specialized trucks and lifts are typically

City Touch: Network "Connected" LA Lights

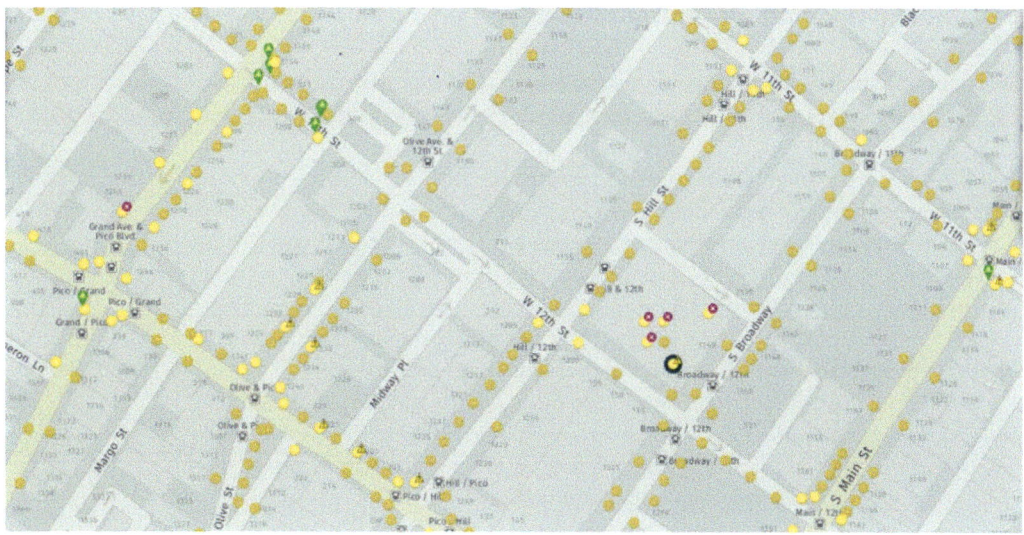

Key: *Bright Yellow: Active*
Green: *Pending Connection*
Dull Yellow: *No node*

Magenta: *Error (light out)*
Yellow Triangle: *Warning (no communication)*

▲ Figure 4.1

City of Los Angeles Street Light Maintenance Geographic Information System. Map dot colors indicate maintenance condition of street lights. Courtesy of Los Angeles Bureau of Street Lighting.

needed. New lighting may increase inspection needs and even trigger citizen complaints that require city staff time to address.

The sheer data management effort for tracking and responding to repair needs is substantial, but may benefit from a Geographic Information System (GIS) application. (See Figure 4.1 for an example of the Los Angeles GIS system screens for maintenance status.)

4.2.3 Life Cycle Costs

Life cycle cost estimates are used to consider both the one-time capital costs and the recurring operations and maintenance costs of a physical improvement or project. Life cycle costs are typically stated as the lifetime cost, combining the initial installation plus the present value of cumulative operations and maintenance costs, including replacement parts. The number of years assumed for the "lifetime" varies, but a typical forecast period is 30 years. LED luminaires generally last 10–20 years, two to four times the life of High Pressure Sodium (HPS) sources.[13] "Payback period" may be used to determine when the net cumulative savings from an improvement will exceed the installation costs. There is the obvious trade-off between the benefits and costs of higher levels of illumination, but also a point of diminishing or even vanishing safety returns is commonly noted as illumination is increased.

4.3 Energy Use and Greenhouse Gas Impacts

The U.S. Department of Energy estimated that in 2017 lighting consumed six percent of all energy and 16 percent of all electricity used in the U.S.[14] Of this total, 6 percent was used by street and area lighting. (The largest single energy user "lighting submarket" was linear indoor lighting, such as fluorescent tubes, at 31 percent of all lighting electricity use.)

The electricity used in lighting represents a major greenhouse gas (GHG) emissions source. The GHG emissions depend on the local electricity source. Approximately 63 percent of U.S. electricity comes from burning fossil fuels, mostly coal and natural gas. However, this varies substantially by location. Lighting accounts for nearly 6 percent of global CO_2 emissions.[15]

The ongoing conversion of lighting to more energy-efficient LED technology should reduce these impacts substantially. "Smart lighting" technology can further this trend. Solar roadway lighting using LEDs is available to minimize the GHG impacts, but it has had very limited municipal usage.

Life cycle cost estimates are used to compare alternative improvement strategies, based on both capital and operations/maintenance costs.

4.4 Light Pollution

There are three main types of light pollution (or "obtrusive light"): light trespass, glare, and skyglow. (See Figure 4.2.) Light trespass and glare impacts are location-specific impacts on adjacent residents and travelers, while skyglow is a more general problem for skies in metropolitan areas. Overlighting (excessive levels and spread of light, clutter from lighting equipment) is sometimes considered the fourth type of light pollution, but it could also be judged a factor often contributing to the other types of obtrusive light. It also is harder to identify and measure than the other categories.

4.4.1 Light Trespass

Light trespass, or "spill light," refers to unwanted illumination of properties adjacent to the roadway, such as private residences. Spill light is disliked especially for potential interference with residents' sleep and privacy. A high "Back" lighting value in the BUG rating (as described in Chapter 2) can be associated with light trespass.

Spill light into private residences should be absolutely minimized. Other facilities, such as a park or office building, are less sensitive.

Spill light can be controlled through physical shielding and careful placement of fixtures. Dimming lights during low-activity periods also may reduce complaints. LEDs generally provide a more precise, vertical beam with less spill than high-intensity discharge (HID) lighting.

4.4.2 Glare

Glare is unwelcome light in the traveler's eye that produces discomfort or disability. It can affect pedestrians, drivers, or bicyclists. Glare metrics were defined in Chapter 2.

Prominent sources of glare include vehicle headlights, street lights, and commercial/residential site lighting. The key factors affecting discomfort glare

| Light Trespass | Skyglow | Glare |

▲ Figure 4.2

Types of Light Pollution. Light pollution is often called "obtrusive lighting." Courtesy of Adobe Stock Photos. © Annafranca - stock.adobe.com. © maislam - stock.adobe.com. © xyzt - stock.adobe.com.

are the luminance of the glare source, the background luminance, the solid angle of the glare source to the subject, and the glare source position in the field of view.[16] Glare is typically produced by light source luminance at 60–90 degrees up from vertical. LEDs generally cause more glare than older HID sources. Higher differentials between the glare source and background luminance cause more glare. Glare sources that are central in the field of view (close to the viewer's fixation point) are also more troubling. Furthermore, sources with higher correlated color temperatures (CCT) and blue-content spectral power distribution (SPD) tend to produce more glare. In fact, the City of Los Angeles found that cooler (higher) CCT values for LEDs generated more resident complaints about glare.[17]

Glare was identified as the most important factor in determining pedestrian satisfaction with different outdoor lighting installations in a survey of pedestrians on a college campus and a small town.[18] Drivers have shielding from their autos, while more exposed pedestrians typically scan a broader area and thus are much more susceptible to glare from lights to their sides.

A lab study found the primary factors affecting perceptions of discomfort glare include:

- Luminance of the glare source
- Distance of the glare source to viewer
- Position in the field of view
- Luminance of the background[19]

Glare can be reduced by the type of lighting and by careful lighting design. IES *Recommended Practice* RP-8 provides standards for veiling luminance ratio for midblock segments but not intersections. IES recommends using luminaires with low output in higher angle range (60–90 degrees from vertical) to reduce glare in intersections.

The physical layout of streets and intersections can also affect glare. For example, median islands can reduce glare in drivers' eyes from oncoming headlamps. Approach grades at intersections can also affect headlamp glare, but they are difficult to retrofit.

4.4.3 Skyglow

Skyglow is unwanted light scattered into the skies. It has gradually been increasing in the developed world for decades. (See Figure 4.3.) It is considered a significant barrier to astronomy and informal stargazing. Accordingly, it has led to the formation of the International Dark-Sky Association and actions by astronomic observatories. Potential impacts on human and animal health and behavior are also a concern.[20]

The different forms of light pollution (light trespass, glare, skyglow) can be substantially mitigated by similar measures, such as shielding and dimming. The choice of the light source and illumination levels also affect light pollution.

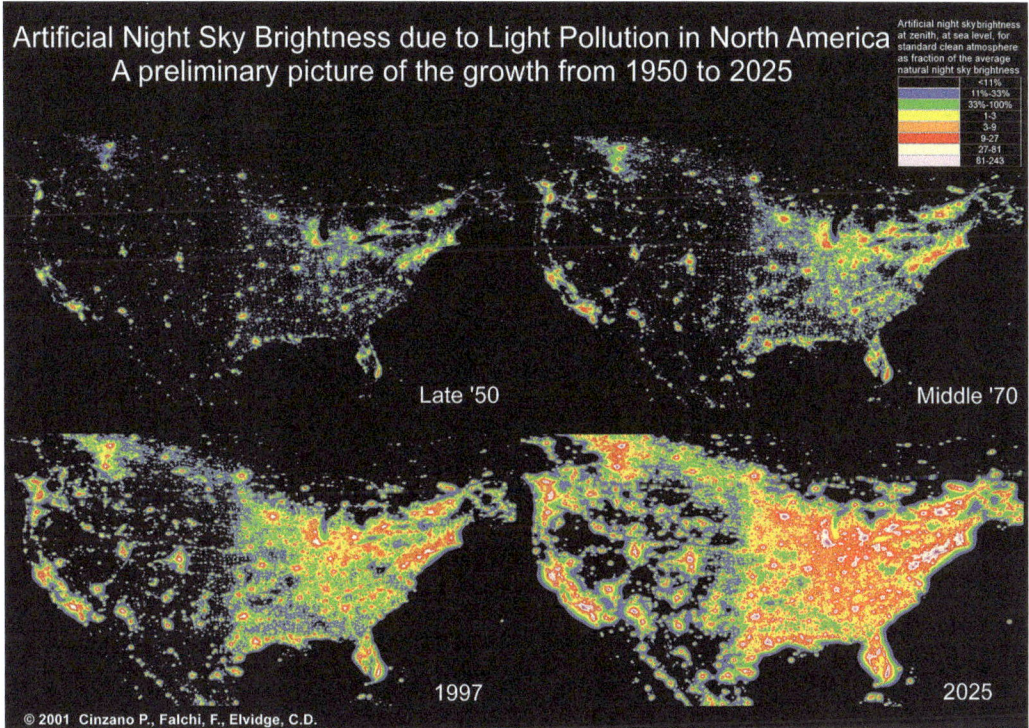

Artificial Night Sky Brightness due to Light Pollution in North America
A preliminary picture of the growth from 1950 to 2025

Artificial night sky brightness at zenith, at sea level, for standard clean atmosphere as fraction of the average natural night sky brightness

<11%
11%-33%
33%-100%
1-3
3-9
9-27
27-81
81-243

Late '50   Middle '70

1997   2025

© 2001 Cinzano P., Falchi, F., Elvidge, C.D.

▲ Figure 4.3

Increasing Skyglow over the U.S. over Recent Decades. The maps are color-coded according to night sky artificial light brightness as a percentage of natural night brightness. The maps transition from 1950, with roughly a dozen metropolitan areas at artificial light brighter than natural night brightness, to a 2025 forecast, with about a third of the country at that level. Courtesy of Fabio Falchi, photographer.

Skyglow is caused by a broad range of lighting, including street lights, private outdoor lighting, indoor lighting, and vehicle headlamps. Skyglow can be produced by light directed downward but reflecting off snow or pavement, not just by light at a high angle. Blue light scatters more than amber light, and this accounts for the blue appearance of the sky.[21]

Skyglow can be reduced by shielding and dimming, as well as by managing illumination levels and selection of specific lighting sources. LED sources, particularly with higher CCT, have greater light scatter and skyglow, but this is often offset by decreased total lumens and lower light trespass to achieve desired visibility levels.[22] The International Dark-Sky Association awards its Fixture Seal of Approval to lighting with CCT of 3,000 degrees Kelvin maximum (down from 4,100 Kelvin), although the organization indicated it may go even lower with future technological advances.[23]

4.5 Human Health Impacts of Lighting

The human species evolved for millennia without artificial lighting. The daily cycle of daytime sunlight and nighttime moon and star light regulating activity and sleep continued until artificial lighting became common in developed nations in the last century.

Organizations and individuals have expressed concerns about health impacts of nighttime lighting, particularly on sleep, but also on obesity and cancer. Some have warned about the contribution of lighting to a general increase in round-the-clock sensory stimulation, along with other modern sources such as televisions, computers, and smartphones.[24] The scientific basis of such concerns should be carefully evaluated. Potential adverse health impacts should also be balanced against the positive physical and mental health benefits of lighting encouraging more exercise and socializing after dark.

4.5.1 Melatonin Suppression

Melatonin is a hormone that helps regulate circadian rhythms and sleep. The American Medical Association (AMA) in 2016 issued a statement expressing concerns about the increasing use of LEDs for outdoor lighting and the impacts on melatonin levels.[25] The AMA stated that

> blue-rich LED streetlights operate at a wavelength that most adversely suppresses melatonin during night. It is estimated that white LED lamps have five times greater impact on circadian sleep rhythms than conventional street lamps. Recent large surveys found that brighter residential nighttime lighting is associated with reduced sleep times, dissatisfaction with sleep quality, excessive sleepiness, impaired daytime functioning and obesity.

4.5.2 AMA Recommendation and Visibility Research

The AMA supported the "proper conversion" to community-based LED lighting to achieve energy and environmental benefits. However, it encouraged minimizing "blue-rich" lighting by use of CCT of 3,000 K or below LEDs for outdoor lighting, with shielding and dimming.

Lighting researchers have argued that CCT is not a definitive metric to determine blue spectral content. CCT gives a single number on the color *appearance* of a light source, but SPD more precisely describes the blue content.

Roadway lighting melatonin suppression is typically much less than objects like a smartphone, tablet, or bedside lamp. (See Figure 4.4.) Research subjects measured by the Virginia Tech Transportation Institute showed much higher illumination falling on the corneas of LED computer watchers (100 lux) and e-reader viewers (31.7 lux) than drivers near 4,000 K LED street lights (1.8 lux).[26] The *Solid-State Roadway Lighting Design Guide* notes that: "Research does not currently show health impacts from properly designed roadway lighting" including 3,000 or 4,000 K CCT LED sources, using recommended lighting levels that meet glare limits and light trespass values.[27]

The Lighting Research Center at Rensselaer Polytechnic Institute also issued a statement criticizing the AMA report for lack of scientific rigor and a "misapplication of metrics" in focusing on the CCT of LED lighting.[28] It pointed out that there are numerous light source factors affecting melatonin levels and

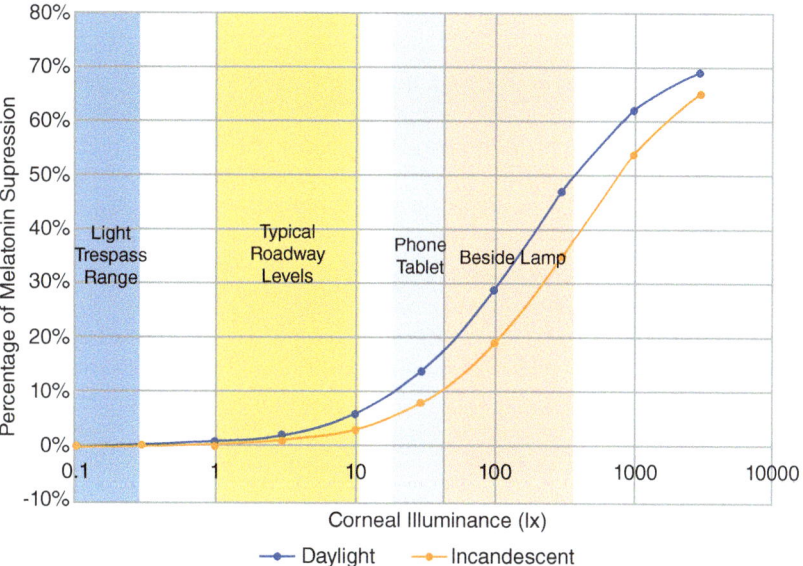

Source: Figueiro et al. (2006) and Kinzey (2016)

circadian rhythms, including the intensity, duration, and timing of exposure, as well as SPD. It pointed to the wide range of relevant light sources besides roadway lighting and the social benefits of lighting and LED conversion.

4.5.3 Other Human Health Impacts

Potential health impacts of artificial light on cancer and obesity have become a concern to some, although research is very limited. Lighting has been implicated in breast cancer in epidemiological studies, with a plausible physiological link. For example, female night shift workers were found at greater risk for breast cancer than daytime female workers.[29] An Israeli study found that women living in neighborhoods bright enough "to read a book at midnight outside" had a 73 percent greater risk of developing breast cancer than those residing in areas with the least outdoor lighting.[30] Whether this correlation was caused by lighting differences is questionable, but the suppression of melatonin has been linked to tumor growth in a number of studies.[31] As discussed above, the melatonin suppression of devices or indoor lighting appears much greater than that due to street lighting.

Researchers in the Netherlands suggested based on studies in mice that prolonged exposure to artificial light may contribute to obesity by inhibiting the fat-burning processes that normally occur during darkness.[32] This was consistent with Israeli research identifying a correlation between outdoor lighting levels and obesity by country. For both the obesity and cancer findings, there are numerous possible confounding factors besides outdoor lighting.

4.6 Flora and Fauna Impacts of Light

The natural daylight cycle shapes the behavior of animals and plants and has through eons of evolution contributed to biodiversity and environmental balance.[33] The phases of the moon are also important for triggering hunting and migration activity.

Light pollution disrupts nocturnal animal behavior and circadian rhythms of humans, animals, and plants. Artificial lighting causes significant harm to many species. For example, in New York City, up to 230,000 birds annually are killed by flying into buildings, in part due to disorientation caused by lighting.[34] Sea turtles hatched on beach sand often instinctively head for lights, apparently an evolutionary strategy that encouraged them to seek water (with moon and stars reflected on it). Bright artificial lighting may lead them away from water and toward hazardous developed areas. (See Figure 4.5.) This results in millions of deaths annually in Florida alone.[35]

Flora are also affected by outdoor lighting, with economic, ecosystem, and aesthetic impacts. For example, soybean growth can be reduced by 20 to 40 percent due to roadside lighting delaying flowering and ripening.[36] One publication lists 65 trees and shrubs that can be adversely impacted by artificial light.[37]

Mitigation is possible, through effective regulations and guidelines, lighting changes, or other methods. For example, in the Netherlands, street lighting

> Light pollution can interfere with animal and plant behavior patterns tied to the natural daylight cycle. These impacts can be mitigated either through changes in lighting or in some cases through non-lighting measures, such as building design.

◄ Figure 4.5

Disoriented Sea Turtles Attracted to Lights. Courtesy of Dawn Witherington, Illustration & Design.

has been specified to refine the spectral content of LED lights to reduce impact upon bat species.[38] There, in Zuidhoek-Nieuwkoop, home to a rare bat species, red frequencies are emphasized in street lights, since these frequencies do not discourage hunting.

As an example of non-lighting mitigation, a recently passed New York City ordinance requires bird-friendly glass (often ultraviolet-coated to be more visible to birds) on the lowest 75 feet of new and renovated buildings. In some cases, bird deaths are related to migration, so this can also be mitigated by switching lights off or dimming them in those seasons, as has been done under a voluntary program in Philadelphia and other cities.[39]

Mitigation measures need to be tailored to sensitive areas, such as near wetlands or wildlife preserves. This requires studying the presence of sensitive species in areas of major planned lighting improvements.

4.7 Fixed Object Hazards

Lighting poles located near the roadway pose a small, but not trivial, safety risk to motor vehicle drivers and passengers. About 20 percent of motor vehicle crash deaths result from a vehicle leaving the roadway and hitting a fixed object alongside the road.[40] Of these fixed-object deaths, 12 percent were due to collisions with utility poles.

Poles also can be a barrier to those with mobility issues. For example, blind and visually impaired pedestrians need to detect the poles. Wheelchair and walker users may be impeded by the poles.

To mitigate fixed-object crashes, the Federal Highway Administration and the American Association of State Highway and Transportation Officials suggest an approach in this priority order:

1. Keeping the vehicle on the roadway
2. Helping the vehicle return to the roadway before hitting an object
3. Reducing the impact if the vehicle hits the object, primarily by using breakaway poles.[41]

Naturally, fewer poles and/or locations further from the curb line reduce the risk of a collision. Energy-absorbing poles are another option. These are designed to wrap around the vehicles upon typical impact force.

Light poles can be significant risks for motor vehicle crashes, impediments to pedestrians with disabilities, and sometimes unattractive streetscape features.

Fewer poles reduce these impacts, but pole placement and design also can mitigate these consequences.

4.8 Aesthetics of Poles and Wires

Street light poles and overhead wires are often considered unattractive features that mar the landscape. Pole height and placement are key factors in aesthetic quality.[42] The relationship of the mounting height to adjacent buildings and street furniture affects street light aesthetic perceptions. Shorter, pedestrian-scale lights appear less intrusive.

Some pole and luminaire types are generally considered more attractive. For example, historic light pole designs often add to the sense of place, as described in the previous chapter. Consistency with street furniture styles, such as benches and planters, can be considered.

Most urban street lights receive power from underground wires, but luminaires mounted on utility poles or in more rural areas often use overhead wiring. Overhead wiring can be undergrounded. However, undergrounding costs are very high, and funding for undergrounding is quite limited.

4.9 Regulatory Tools for Controlling the Adverse Impacts of Lighting Projects

The nighttime pedestrian environment along the sidewalk or path is often affected by standard lighting for buildings, plazas, and other adjacent properties. Lighting for private billboards and signs is also important in certain locations. The adjacent site and building lighting can add visual interest, increase illumination levels on the sidewalk, and help in the recognition of places of interest at night. However, it can also cause excessive shadows, distraction, and glare.

The following is a short review of **tools for regulating private lighting.** This topic is addressed in more detail in other publications, such as Illuminating Engineering Society (IES) RP-39-19 *Recommended Practice: Off-Roadway Sign Luminance* (for internally lit signs) and IES RP-33-11 *Recommended Practice: Lighting for Exterior Environments.*[43]

Private lighting adjacent to the street right-of-way is more difficult for government or advocates to control than public lighting. Private building, sign, and site lighting can be influenced through such mechanisms as lighting ordinances, sign ordinances, architectural reviews, and environmental reviews. Lighting ordinances are a more direct, transparent, consistent, and powerful measure, while the other tools are indirect, often opaque, and generally less consistent. All these measures focus on controlling adverse impacts, rather than promoting positive qualities. To varying degrees, these tools (especially environmental review) also can influence public lighting.

> Private building and site lighting frequently impact the pedestrian realm. Regulatory tools for controlling building and site lighting impacts include lighting ordinances, sign ordinances, environmental reviews, and architectural reviews.

The IES/International Dark-Sky Association Model **Lighting Ordinance** is aimed at most outdoor lighting except street and roadway lighting.[44] Its goals are to:

- Limit the amount of light used
- Minimize glare
- Minimize skyglow
- Minimize light trespass

Thousands of communities have adopted lighting ordinances.[45] Some states like New York and California have legislated similar provisions. Local governments that adopt the model ordinance may use either a prescriptive method or a performance method for determining whether a property or site complies. The prescriptive method includes a total site lumen limit, limits on luminaire BUG (Backlight/Uplight/Glare) levels, and light shielding for parking lot illumination. The performance method includes a total site lumen limit and limits on lumens and vertical illuminance at the property boundary. The model ordinance considers the lighting zone of the area. Lighting zones range from LZ0 (with no ambient lighting) to LZ4 (high ambient lighting). Some cities, such as Plymouth (Minnesota) and New York City have formally adopted lighting zones. New York City ties lighting zones directly to zoning classifications, with LZ0 for parks and LZ4 for high-density commercial districts.[46]

Some **sign ordinances**, such as that for Arlington County, Virginia, specifically address lighting related to signs. The Arlington ordinance, a zoning ordinance section, has extensive controls on lighted signs, both internally and externally illuminated.[47] It includes maximum allowable illuminances (based on zoning, location, and type of sign), hours of illumination permitted, and requirements for shielding and dimming. It also prohibits projected signs, searchlights, flashing signs, reflective signs, or those that are "excessively bright" as determined by the Zoning Administrator.

A fairly small number of cities use an **architectural review** board (ARB) to evaluate and possibly regulate building design. For example, the Norfolk, Virginia ARB reviews exterior alterations and construction, with strong regulatory powers in historic districts.[48] The ARB requires submittal of floors plans and building elevations with details on lighting design, along with separate narrative information on proposed lighting.

Under the federal **environmental review** law (NEPA, the National Environmental Protection Act), lighting for federally controlled, funded, or permitted projects can be found to cause significant adverse visual impacts.[49] Impacts can include degraded views of a scenic resource, like an historic building or landmark tree.[50] Sources of impact could include construction lighting, physical lighting features, and operations and maintenance activities. Mitigation measures may include: limits on construction lighting, landscaping as a buffer, limits on light pollution per Illuminating Engineering Society (IES) and International Dark-Skies Association (IDA) guidance, or ornamental lighting. Guidelines that cover environmental documents such as Environmental Impact Statements suggest that for some highway projects, technical studies of lighting may be required. Photo-simulations of lighting and landscaping plans are helpful to convey impacts to readers.

The California Environmental Quality Act (CEQA) applies to a broad range of projects reviewed or sponsored by local and state governments within that state, and some other states have similar laws. The CEQA official checklist includes a potential finding that a real estate development or infrastructure project could create a significant adverse environmental impact by "a new source of substantial light or glare that would adversely affect daytime or nighttime views of the area."[51] It is also common for Environmental Impact Reports for major California projects to address visual impacts of the physical project in detail (e.g., with photo-simulations of new buildings). So there is the potential to include a review of the aesthetics of illumination and light equipment. CEQA also requires identifying impacts to historic resources, which can include street lights. The San Francisco Better Market Street project, for example, included extensive consideration of potential impacts to historic "Path of Gold" street lights in its Environmental Impact Report.[52]

Notes

1. Ronald Gibbons, "Connected Infrastructure Activities," *Pedestrian and Bicycle Information Center Webinar on Lighting for Pedestrian Safety and Walkability*, October 17, 2018, https://www.pedbikeinfo.org/webinars/webinar_details. cfm?id=13.
2. University of North Carolina Highway Safety Research Center, Vanasse Hangen Brustlin Inc., and Toole Design Group, *PEDSAFE: Pedestrian Safety Guide and Countermeasure Selection System* (Washington, DC: FHWA, 2013), http://www. pedbikesafe.org/pedsafe/countermeasures_detail.cfm?CM_NUM=8.
3. Michael Kato, City of San Mateo Public Works Department, Email, April 6, 2021.
4. San Francisco Municipal Transportation Agency, *Balboa Park Station Capacity and Conceptual Engineering Study* (San Francisco: SFMTA, 2012), https://www.sfmta. com/sites/default/files/agendaitems/11-6-12item11balboaparkphaseii-revisedpublic draftreportoct2012accessibleforboard.pdf.
5. Brookline (Massachusetts), "Draft Report: Extending Pedestrian-Friendly Lighting," May 2020, https://www.brooklinema.gov/DocumentCenter/View/22072/Pedestrian-Friendly-Street-Lighting-Committee-Draft-Report.
6. Alameda County (California) Transportation Commission, Bicycle and Pedestrian Master Plan Cost Estimating Tool, 2019.
7. Max Bushell et al., *Costs for Pedestrian and Bicyclist Infrastructure Improvements: A Resource for Researchers, Engineers, Planners and the General Public* (Chapel Hill: University of North Carolina Highway Safety Research Center, October 2013), http://www.pedbikeinfo.org/cms/downloads/Countermeasure%20Costs_Report_ Nov2013.pdf.
8. Bushell et al., *Costs for Pedestrian and Bicyclist Infrastructure Improvements*.
9. Los Angeles Bureau of Street Lighting, *LA Lights Strategic Plan 2020–2025* (Los Angeles: BSL, 2020), http://bsl.lacity.org/strategic_plan.html.
10. Gabe Arnold and Brian Buckley, *LED Street Lighting Assessment and Strategies for the Northeast and Mid-Atlantic* (Lexington, MA: Northeast Energy Efficiency Partnerships, January 2015), https://neep.org/sites/default/files/resources/

DOE_LED%20Street%20Lighting%20Assessment%20and%20Strategies%20for%20
the%20Northeast%20and%20Mid-Atlantic_1-27-15.pdf.

11. LABSL, *LA Lights Strategic Plan 2020–2025*.

12. "Phoenix Completes LED Street Light Conversion, Estimating $3.5 Million in Energy
Savings Per Year," City of Phoenix website, accessed March 7, 2021, https://www.
phoenix.gov/news/street-transportation/2481.

13. Matt A. V. Chaban, "LED Streetlights in Brooklyn Are Saving Energy but Exhausting
Residents," *The New York Times*, March 23, 2015, https://www.nytimes.
com/2015/03/24/nyregion/new-led-streetlights-shine-too-brightly-for-some-in-
brooklyn.html.

14. Navigant Consulting for the US Department of Energy, *Energy Savings Forecast of
Sold-State Lighting in General Illumination Applications* (Washington, DC: DOE, 2019).

15. "LED," The Climate Group website, accessed March 7, 2021, https://www.
theclimategroup.org/led.

16. Yulia Tyukhova, "Discomfort Glare in Outdoor Nighttime Environments," IES
Webinar, June 18, 2020.

17. Los Angeles Bureau of Lighting, "Changing Our Glow for Efficiency," Municipal
Solid State Lighting Consortium LED Workshop, Los Angeles, April 2012.

18. University of Washington Integrated Design Lab, *Campus Illumination: A Roadmap
to Sustainable Exterior Lighting at the University of Washington Seattle Campus*
(Seattle: University of Washington, 2017), https://www.lightingdesignlab.com/sites/
default/files/pdf/Campus-Illumination-Roadmap-final.pdf.

19. Tyukhova, "Discomfort Glare in Outdoor Nighttime Environment."

20. Lucy K. McLay et al., "What Is the Available Evidence That Artificial Light at Night
Affects Animal Behaviour? A Systematic Map Protocol," *Environmental Evidence* 8,
no. 7 (2019), doi: 10.1186/s13750-019-0151-9.

21. Gibbons, "Connected Infrastructure Activities," 2018; Ian Ashdown, "Light Pollution
Depends on the Light Source CCT," *LEDs Magazine*, October 20, 2015, https://
www.ledsmagazine.com/smart-lighting-iot/white-point-tuning/article/16695938/
light-pollution-depends-on-the-light-source-cct-magazine.

22. National Academies of Science, Engineering and Medicine, *Solid-State Roadway
Lighting Design Guide. Volume 1: Guidance* (Washington, DC: The National
Academies Press, 2020), doi: 10.17226/25678.

23. "Fixture Seal of Approval," International Dark-Sky Association website, accessed
March 7, 2021, https://www.darksky.org/our-work/lighting/lighting-for-industry/fsa/.

24. Jane Slade, "Nature Leads the Way: What if We Reordered the Design Process?,"
Lighting Design and Application 50, no. 6 (June 2020): 16–18.

25. "AMA Adopts Guidance to Reduce Harm from High Intensity Street Lights,"
American Medical Association website, accessed March 7, 2021, https://www-
ama-assn-org.ezp-prod1.hul.harvard.edu/press-center/press-releases/ama-adopts-
guidance-reduce-harm-high-intensity-street-lights.

26. Ray Bhagavathula, "Lighting and Health," IES Street and Area Lighting Conference,
Dallas, Texas, October 28, 2020.

27. National Academies of Science, Engineering and Medicine, *Solid-State Roadway
Lighting Design Guide. Volume 1: Guidance* (Washington, DC: The National
Academies Press, 2020), doi: 10.17226/25678.

28. Mark S. Rea and Mariana G. Figueiro, "Response to the 2016 AMA Report on LED
Lighting," Rensselaer Polytechnic Institute – Lighting Research Center website, June
30, 2016, https://www.lrc.rpi.edu/resources/newsroom/AMA.pdf.

29. Sarah P. Megdal et al., "Night Work and Breast Cancer Risk: A Systematic Review,"
European Journal of Cancer 41 (2005): 2023–2032.

30. Ron Cepesiuk, "Missing the Dark: Health Effects of Light Pollution," *Environmental Health Perspectives* 117, no. 1 (January 2009): A20–A27, https://www.ncbi.nlm.nih.gov/pmc/articles/PMC2627884/.

31. Samar A. Jasser, David E. Blask, and George C. Brainard, "Light during Darkness and Cancer: Relationships in Circadian Photoreception and Tumor Biology," *Cancer Cause Control* 17 (2006): 513–523; Eva S. Schernhammer et al., "Night Work and Risk of Breast Cancer," *Epidemiology* 17 (2006): 108–111.

32. Nicole Makris, "Artificial Light Associated with Obesity, Study Says," *Healthline*, August 1, 2019, https://www.healthline.com/health-news/artificial-light-associated-with-obesity-study-says-051115#1.

33. Slade, "Nature Leads the Way."

34. Slade, "Nature Leads the Way."

35. Brianna Lynne, "Tourism Can Worsen Deadly Light Pollution in Sea Turtle Habitats," Earth.com News website, December 21, 2019, https://www.earth.com/news/tourism-light-pollution-sea-turtle/.

36. Chi Ling Chen et al., "Effect of Night Illumination on Growth and Yield of Soybean," *Journal of Taiwan Agricultural Research* 58, no. 2 (2009): 146–154.

37. William R. Chaney, *Does Night Lighting Harm Trees?* Circular FNR-FAQ-17 (W. Lafayette, IN: Department of Forestry and Natural Resources, Purdue University, 2002).

38. "Tourism Can Worsen Deadly Light Pollution in Sea Turtle Habitats," earth.com, accessed March 7, 2021, https://www.earth.com/news/tourism-light-pollution-sea-turtle/.

39. Associated Press, "Philadelphia Calls for 'Lights Out' after Skyscrapers Cause Hundreds of Bird Deaths," *The Guardian* website, March 12, 2021, https://www.theguardian.com/us-news/2021/mar/12/philadelphia-birds-skyscrapers-deaths-lights-out.

40. "Fatality Facts 2019: Collisions with Fixed Objects and Animals," Insurance Institute for Highway Safety website, March 2021, https://www.iihs.org/topics/fatality-statistics/detail/collisions-with-fixed-objects-and-animals.

41. "Safety: Noteworthy Practices: Roadside Tree and Utility Pole Maintenance," US Department of Transportation, Federal Highway Administration website, February 16, 2017, https://safety.fhwa.dot.gov/roadway_dept/countermeasures/safe_recovery/clear_zones/fhwasa16043/ch1.cfm.

42. Illuminating Engineering Society (IES), *RP 8-18: Recommended Practice for Design and Maintenance of Roadway and Parking Facilities Lighting* (New York: IES, 2018), 2–9; Ronald Gibbons, "Connected Infrastructure Activities," webinar.

43. Illuminating Engineering Society, RP-39-19 *Recommended Practice: Off-Roadway Sign Luminance* (New York: IES, 2019); IES, RP-33-11 *Recommended Practice: Lighting for Exterior Environments* (New York: IES, 2011).

44. Illuminating Engineering Society and International Dark-Sky Association, Model Lighting Ordinance with User's Guide, June 2011, https://www.darksky.org/wp-content/uploads/bsk-pdf-manager/16_MLO_FINAL_JUNE2011.PDF.

45. Lighting Ordinances," International Dark-Sky Association website, accessed December 24, 2020, https://www.darksky.org/our-work/lighting/public-policy/lighting-ordinances/.

46. Jack Bailey, "Speaking in Code: Demystifying the 2016 New York Energy Code," Slideshare, April 17, 2016, https://www.slideshare.net/LEDucationNYC/speaking-in-code-demystifying-the-2016-new-york-energy-code.

47. Duncan Associates for Arlington County, Virginia, "Arlington County Sign Ordinance," July 24, 2012, https://building.arlingtonva.us/wp-content/uploads/sites/38/2016/08/ACZOSection34.pdf.

48. City of Norfolk, Virginia, "Architectural Review Board," website, accessed October 24, 2020, https://www.norfolk.gov/1090/Architectural-Review-Board.

49. ICF International, *Guidelines for Visual Impact Assessment of Highway Projects* (Washington, DC: Federal Highway Administration, 2013), https://www.environment.fhwa.dot.gov/env_topics/other_topics/VIA_Guidelines_for_Highway_Projects.aspx.

50. Caltrans (CA Department of Transportation), "Chapter 27: Visual and Aesthetics Review," *Standard Environmental Reference* (Sacramento, CA: Caltrans, 2020), https://dot.ca.gov/programs/environmental-analysis/standard-environmental-reference-ser/volume-1-guidance-for-compliance/ch-27-visual-aesthetics-review.

51. Association of Environmental Professionals, *2019 CEQA Statute and Guidelines* (Palm Desert, CA: AEP, 2019), 313, https://resources.ca.gov/CNRALegacyFiles/ceqa/docs/2019_CEQA_Statutes_and_Guidelines.pdf.

52. San Francisco Planning Department, *Better Market Street Project: Draft Environmental Impact Report, Case No. 2014.0012E* (San Francisco: Planning Department, 2019), 4–8.

Chapter 5

Basic Options in Lighting Equipment

5.1 Purpose and Scope of This Chapter

This chapter describes street lighting equipment that is widely used. Luminaires (light fixtures) and poles are critical determinants of light levels and distribution, adverse impacts, and aesthetic effects. Thus, an understanding of lighting equipment options is a valuable initial step in improving the pedestrian realm.

In U.S. practice, street lights are routinely used on urban arterial and collector streets, most commercial streets, as well as many local residential streets. Illuminating Engineering Society (IES) *Recommended Practice* RP-8 suggests that roadway lighting is less valuable on freeways, rural highways, and local residential streets, where vehicle headlights may be sufficient and vehicle-pedestrian conflicts less likely.

This chapter first compares conventional street lights used primarily for roadway lighting with pedestrian-scale lighting intended especially to illuminate walkways. ("Pedestrian-scale lighting" is not a precise, technical term, and the dividing line between roadway-scale lighting and pedestrian-scale lighting is not clear-cut.) Next, this chapter addresses lighting source technology. In recent years, light-emitting diode (LED) technology has become the overwhelming choice for new installations by cities. LED retrofits for street lighting have also become routine.

Finally, non-lighting measures for improving after-dark pedestrian visibility are addressed. These measures include high visibility roadway markings and reflective clothing.

Earlier chapters provided important background regarding the technical terms and concepts used in this chapter, as well as the potential positive and negative impacts of lighting equipment choices. Chapter 6 will next address innovative technologies, some off-the-shelf and others that are custom treatments, including "smart lighting" options facilitated by LEDs. Chapter 7 includes examples of municipal plans with guidelines for installing both roadway lighting and pedestrian-scale lighting. Final Chapter 10 covers possible technological advances to lighting equipment in the more distant future.

 DOI: 10.4324/9781003149750-5

5.2 Conventional Roadway Lighting

Conventional roadway lighting is intended primarily to light the roadway and secondarily any adjacent sidewalk. Luminaires are mounted over 20 feet (6 meters) high, often about 30 feet (9 meters) high, with poles usually spaced at least 100 feet (30 meters) apart. (See Figure 5.1 for a visual comparison of conventional cobra head-style roadway-scale lighting and pedestrian-scale lighting.)

Desired illumination levels and distribution (for roadway, crosswalks, and sidewalks) are the main **considerations in selecting specific poles and luminaires** and designing the pole layout. Illumination levels from a new street light on pedestrians are determined mainly by the light source technology

The lighting designer selects the equipment (poles and luminaires) and the pole layout, aiming in part to achieve the desired illumination levels for roadways, crosswalks, crosswalk entry points, and sidewalks. Pedestrian volumes and crossing locations are major factors in determining where to locate street lights.

▶ Figure 5.1

Cobrahead Roadway Light Next to Pedestrian-Scale Light.

(e.g., LED), photometric characteristics, the distance light travels to the pedestrian (mounting height and spacing), the reflectivity and brightness (luminance) of the pavement, interference from foliage, and the reduction in light output due to aging or maintenance issues. (Over time, a luminaire's effectiveness may be reduced as parts wear out and dirt gathers on the lens.)

Designers need to match equipment to the project needs, including local policies, client preferences, and budget, as well as location characteristics. Special site considerations may include the collision history, adjacent land uses, trees, weather, and shadows.

The lighting designer determines key lighting features such as:

- Light type and mounting
- Lighting source technology
- Light loss factor
- Pole type, height, and luminaire arm length
- Pole offsets and spacing
- Luminaire shape and other characteristics
- Luminaire wattage and output
- Luminaire optical distribution and BUG (Backlight, Uplight, Glare) rating.

The primary types of **mounting** used for conventional street lights include post-top or arm-mounted. Utility-pole-mounted lights are also common for roadway lighting. Other types of mounting rarely used for roadway lighting include:

- High mast
- Wall-mounted
- In-roadway lighting (to delineate lane/edge lines or crosswalks)

Lighting source technology is discussed in detail below.

Light loss factor indicates the reduction in output with aging, due to parts wearing out, dirt, and similar factors. Designers aim to provide sufficient light output so that even with this gradual reduction, light levels 20 or more years after installation will meet standards or target levels.

Pole height and the length of any arm affect the distribution and uniformity of light, as well as its appearance and attractiveness.

Pole offset pattern and spacing significantly affect the distribution and uniformity of light. Poles may be placed on only one side of the street or both sides with varying patterns. Pole placement can impede walking or wheelchair use, especially by pedestrians with disabilities.

The luminaire includes an enclosure and connection to mounting, an electrical control device, wiring, optical controls, and other control hardware (e.g., fusing, photo controls, monitoring, and dimming controls). Luminaire output (lumens) and wattage (power use) are key considerations in efficacy. The distribution of light can be controlled optically and by physical shielding.

Designers and buyers often review manufacturer information in catalogs, spec sheets, photos, photometry files, and diagrams. Some manufacturers provide the ability to customize a 2D or 3D rendering of a scene with their products, potentially viewable with augmented reality technology. Information is provided on product characteristics such as:

- Light source technology
- Output (lumens)
- Power use (wattage) and efficacy (lumens per watt)
- Light distribution
- Dimming controls
- Mounting type and height
- Luminaire shape
- Correlated Color Temperature (CCT)
- Color rendering

A Pacific Northwest National Laboratory study found there were five primary variables that influenced *pedestrian ratings of outdoor lighting* that the designer should consider:

- Luminaire appearance
- Glare
- Uniform distribution of lighting preferred
- Warmer color appearance (CCT of 2,700–3,000 degrees Kelvin) preferred
- Horizontal illuminance near the lower end of the IES recommended range preferred if glare was kept low[1]

However, it should be noted that the study was conducted on the Stanford University campus and a very small town (Chappaqua, New York), so it is questionable how applicable it is to most urban street settings.

5.3 Pedestrian-Scale Lights

Pedestrian-scale lighting is intended primarily to light the sidewalk and cross-walk entry points, or separated walkways (such as a park path). It is used especially for the security and comfort of pedestrians, including orientation and wayfinding, as well as safety from tripping and slipping hazards. On some narrow streets, or where mounted on both sides of the street, pedestrian-scale lights can also illuminate the roadway sufficiently so that no roadway-scale lighting is needed. Luminaires are typically post-top mounted, lower than roadway-scale lights, usually 10 to 18 feet (3–5 meters) high, with poles spaced closely, often about 50 feet (15 meters) apart. Other less common options for mounting include bollards or (very rarely) in-pavement lights. (There are anecdotal concerns that in-pavement lights have a greater incidence of maintenance issues.)

Pedestrian-scale lights are often considered optional, supplemental additions where roadway lights are already installed on wider streets. (Prioritizing their use on significant pedestrian routes is discussed in Chapter 7.) Pedestrian-

"Pedestrian-scale lighting" is not precisely defined, but it refers to lighting primarily serving walkways. Typically, luminaires are post-top-mounted about 10 to 18 feet high, spaced more closely than conventional roadway-scale lights (roughly 50 feet apart). Bollards or, in rare cases, in-pavement lights are also used.

scale lights are also used alone on separated walkways where significant after-dark use is expected, such as an urban park.

There is limited research available on how much pedestrian-scale lighting (as opposed to new illumination generally) reduces the **risk of pedestrian-involved crashes.** Virginia Tech Transportation Institute researchers found that pedestrian-scale lights (18 feet or 5 meters high) significantly improved pedestrian detection of tripping hazards in the roadway compared to roadway-scale (30-foot or 9 meters) lights.[2] However, roadway-scale lights were slightly better for driver detection of pedestrian mannequins, but not to a statistically significant extent.

For lighting design on urban streets, the **equipment selection considerations** for pedestrian-scale lighting are generally similar to those for conventional street lights described in the previous section. If lighting is only needed for sidewalks on one side, a more narrow distribution of light is desired, hence lower mounting heights and closer spaced poles. Also, the aesthetics of pedestrian-scale poles and luminaires are often considered more important than roadway lighting.

There are special considerations in selecting equipment for off-street use, such as plazas, parking facilities, and park pathways. For example, park pathway lighting may be integrated with lighting for lawns and parking lots. Special lighting may be needed for sculptures or other features. Aesthetic requirements are often a high priority, such as for an unobtrusive or "woodsy" appearance. Light obstruction from a continuous tree canopy may require very low mounting heights. Tree roots may limit electrical conduit and pole placement options.

For a recreational bike/pedestrian path, use may be legally restricted after dark, reducing or eliminating the need for lighting. Obstructions and narrow right-of-way may severely limit options for pole placement. Plaza lighting is used for multiple purposes: to illuminate paths, to highlight special features like sculptures or seating, and sometimes as an accent or artistic feature itself.[3] There are numerous potential mounting and placement options, which may be combined or layered:

- Path lights
- Ceiling and hanging lights
- Wall lights
- Post-top lights
- Wall or column-top lights
- Landscape lights
- Deck and step lights

Plaza lighting should be scaled to the immediate environment. One option for land-scaping lighting is low voltage (12 volts) systems. These use less energy and do not require conduits. Lighting rated for damp or wet conditions should be considered.

5.4 Light Source Technology

Light source technology makes a major difference in how visible pedestrians are to drivers and how pedestrians view their environment. In recent years, LED (Light-Emitting Diodes) lights have become the consensus choice for local governments for new and replacement street lighting, primarily due to their energy and maintenance advantages. In fact, it is questionable whether HID (High-Intensity Discharge) lighting will be readily available for purchase for new installations soon. However, HID lighting, such as High-Pressure Sodium (HPS) or metal halide, is still common, despite numerous LED retrofit projects. Other less common sources for roadway or path lighting include LPS (Low-Pressure Sodium), compact fluorescent, incandescent, and induction lamp.

5.4.1 Description and Benefits of LED Lighting

LEDs are solid-state semiconductor devices that light up when a current flows through them. LED lamps typically use an array of diodes emitting light of different wavelengths (colors). Therefore, lighting is more easily controlled or programmed. For example, lighting can be dimmed, or color appearance and spectral content varied. The LED light source luminance is also more concentrated than in HID, which can cause more glare.

LED lights can be extremely small, durable, and longer lasting than most other options. They have very low power consumption and high efficacy (the amount of useful light output per unit of electricity, expressed as lumens per watt). Modern LED luminaires generally provide greater than 100 lumens per watt.

While LED light can more easily be focused on a smaller area, that may lead to poorer illumination of sidewalks from street lights targeted to light the roadway. Another primary disadvantage is sensitivity to heat. Color rendition is generally better than HID, which provides potential safety and security advantages. Prices have dropped rapidly in recent years to make LED life cycle costs highly competitive.

LED products tend to vary more in lighting characteristics than HID and also have a greater range of capabilities, therefore they are not as simple for cities and utilities to specify and purchase.[4] For example, the light distribution tends to vary more. Color and brightness appearance also vary.

LED luminaires have increased the potential benefits of street lighting for the pedestrian environment, energy use, and "Smart City" capabilities. But they have also complicated the design and procurement of equipment.

HID lamps (including high-pressure sodium, metal halide, and mercury vapor) produce light by exciting gases or metal vapors with electricity in the lamp or tube. They require warm-up and cool-down periods before they can be re-energized. High-pressure sodium, an especially common source, emits a yellowish light from a sodium vapor in an arc tube. Advantages include fairly high efficacy, long life, high light output regardless of positioning of the lamp, a wide range of available wattages, and a good maintenance record. Color rendition, power consumption, and efficacy are often inferior to LEDs.

5.4.2 Routine Installation on Roadways

LEDs are the routine choice for new roadway lighting. In numerous North American cities, such as Los Angeles, San Francisco, New York, Detroit, and Vancouver, HID lighting has been converted to LEDs. Agencies such as the World Bank and the U.S. Department of Energy advocate such conversions for energy and environmental benefits.

Evaluations of LED conversions consistently indicate substantial energy savings, such as the Los Angeles experience described in Chapter 4. The Los Angeles Bureau of Street Lighting (LABSL) also found a perception of improved visibility and higher illumination levels by residents and the Police Department.[5] (See Figure 5.2 for a color rendition comparison of LED and HPS lighting.)

▲ Figure 5.2

Color Rendition Comparison of HPS versus LED Lighting. Before/HPS (upper left) and after/LED (lower right) views from Los Angeles LED Retrofit. Courtesy of City of Los Angeles Bureau of Street Lighting.

A number of other benefits were noted by LABSL, such as full illumination immediately after turn-on and reduced maintenance inventory needs. Lamp failure rate was cut dramatically. The Bureau noted a reduction in crime on affected streets, although it is not clear whether this was due to the retrofit. However, higher (cooler) temperature LEDs generated more public complaints.

The Bureau used an extensive public outreach campaign before converting to LEDs, emphasizing energy and environmental benefits. The campaign included public surveys, a dedicated phone line, and a website.

Chicago has undertaken a similar size LED conversion, replacing 270,000 HPS street lights with connected LED fixtures.[6] This retrofit is expected to save $10 million annually in energy savings and maintenance.

5.4.3 Safety Record

The value of LED lighting for improving pedestrian safety has been indicated by test track research and some field experience. (A large, systematic evaluation of the roadway safety impacts of LED lighting is being conducted by the Pacific Northwest National Laboratory in the Philadelphia region.[7] However, results were not available by publication, and the study does not focus on pedestrian-involved injuries specifically.)

Small target detection distances were statistically significantly greater for LED lighting with "cooler appearing" CCT of 4,100 degrees Kelvin compared to warmer HPS lighting (and to 3,500 degrees LED).[8] In fact, even LEDs dimmed to 25–50 percent of full output generally provided longer detection distances than full-powered HPS lighting. (See Figure 5.3. The colored bars give the mean detection distance. If the thin "error bar" lines at the top for different source/CCT/output combinations do not overlap, there is a statistically significant difference.)

Based on this study and an extensive range of other research, the National Academies *Solid-State Roadway Lighting Design Guide* concluded: "In general, LEDs are said to be better than HPS light sources in terms of photometric and economic performance."[9] The *Guide* also noted "there does appear to be an advantage to using 4,000 Kelvin sources rather than 3,000 or 5,000."[10]

There are also informal indications of the safety benefits of actual LED conversions. For example, Vancouver, British Columbia, implemented an LED conversion, prioritizing locations with higher crash profiles. The City found: "Recent LED street light improvements at high-collision intersections have shown significant and measurable progress made towards the City's zero traffic fatalities goal."[11] LED street lights were installed at 125 signalized intersections, with a

LED lights provide improved visibility and color rendition. However, the more focused beam may lead to lower illumination of sidewalks when they replace HID lighting if light distribution is not carefully considered.

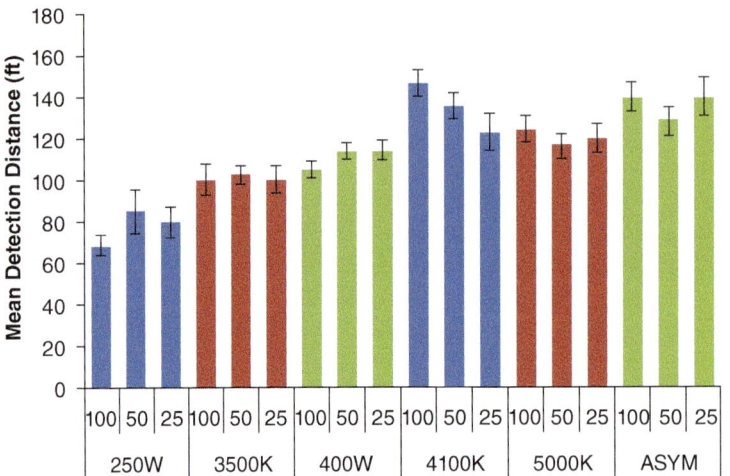

◀ Figure 5.3

Detection Distances for HPS Versus LED Lighting. Bar graph shows the mean distances objects in roadway were detected by observers under different lighting conditions. The light sources compared were 250 watt HPS, 400 watt HPS, 3500 K LED, 4100 K LED, 5000 K LED (all with Type II or oval light distribution), and 5000 K LED with asymmetric (unidirectional) light distribution. For each light source, bars are provided for 100% output, 50% output, and 25% output. If thin error bars at top do not overlap, the differences are statistically significant. Courtesy of Clanton & Associates.

21 percent reduction in total traffic-related fatalities and injuries after dark and a 65 percent reduction in traffic-related fatalities and injuries involving pedestrians after dark. [12] (Data were from 2007 to 2017 period. For each location, the years before and after the installation were averaged. These numbers were compared to a control dataset of 50 intersections that were not modified and that are of similar characteristics to intersections treated. The decreases reported were measured compared to the control dataset. The odds ratio method was used to calculate the safety benefits.) These safety impacts were credited with plans to include the LED conversion in roadway reconstruction and development projects, such as the Burrard Bridge Corridor Project and the Point Grey Seaside Greenway Project.

5.4.4 LEDs May Limit Light Outside Roadway

LED luminaires typically have a more focused light distribution than HID luminaires. Traditional methods of lighting design that focus only on the roadway, assuming that sidewalks would still get adequate spillover lighting from HID lights, may lead to inadequate lighting of sidewalks and bike lanes when LEDs are used.

To counter this, the *Solid-State Roadway Lighting Guide* recommends that illumination on sidewalks and shoulders be checked.[13] Illumination on an area 12 feet (3.6 meters) adjacent to the travel way should be illuminated at 80 percent of the illumination on the nearest travel lane. (That is, the **Surround Ratio** should be 0.8.) This is based partly on VTTI test track research discussed in Chapter 3 showing improved pedestrian detection at higher surround ratios and higher illumination levels. (See Figure 5.4. The graph on the left shows detection distance for drivers observing a pedestrian two feet from the roadway. The graph on the right is based on the pedestrian 10 feet from the roadway.)

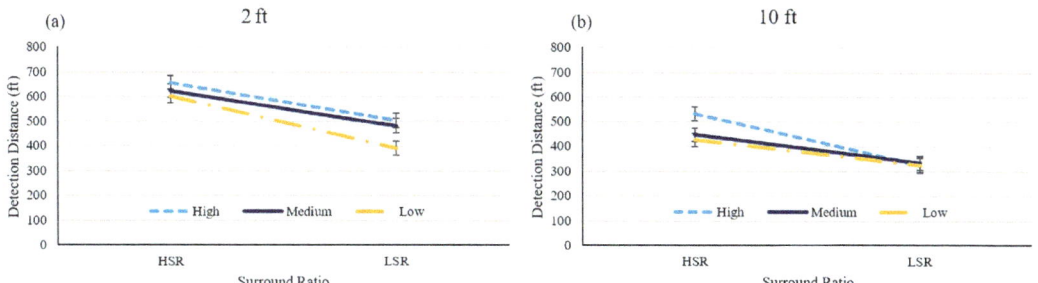

(a) 2 ft   (b) 10 ft

Note: Values are means of detection distances and error bars indicate standard error.

▲ Figure 5.4

Effect of Surround Ratio and Illumination Levels on Pedestrian Detection Distance. Two line graphs both show detection distance to pedestrian (vertical axis) versus decreasing surround ratio (horizontal axis) with lines for high, medium, and low light levels. Left graph shows results for pedestrian offset two feet from driving lane. Right graph is for pedestrian offset ten feet from driving lane. Reproduced with permission from the National Academy of Sciences, Courtesy of the National Academies Press, Washington, DC.

LEDs potentially exacerbate light trespass and glare issues due to their brightness and color appearance. Mounting height and LED placement can contribute to glare. For example, a post-top luminaire with LEDs concentrated on the low end of the luminaire can direct more obtrusive light directly toward residents if not properly shielded.[14]

5.5 Visibility and Safety Supplements to Lighting

There are non-lighting measures that individuals and agencies can use to improve nighttime pedestrian visibility and safety. Light-colored clothing, retroreflective clothing and accessories, flashlights, and retroreflective pavement markings or raised pavement markers can improve visibility of pedestrians or roadways markings.

Lighter shades of clothing are commonly understood to be safer for pedestrians. Research bears this out. For example, a test of driver detection of pedestrians crossing a test track found that pedestrians wearing white clothing even under low vertical illuminance were detected at roughly double the mean distance of pedestrians wearing black or denim (dark blue) clothing.[15] The mean detection distance differences between white and darker clothing were far greater than the differences made by varying vertical illumination levels.

Reflective or retroreflective clothing and accessories make pedestrians more visible to drivers. ("Retroreflective" materials reflect light back toward its source with minimal scattering.) For example, reflective clothing and accessories are often used by recreational runners and cyclists, as well as nighttime outdoor workers. Some pedestrian safety efforts or campaigns, such as the San Francisco PedSafe project, have distributed retroreflective clothing accessories marked with safety slogans.[16] Of course, their use cannot be assumed even after they are provided.

Research has shown visibility benefits for retroreflective clothing (although no research was found directly addressing impacts on pedestrian-involved collisions). Drivers on a test track were better able to recognize a pedestrian on the shoulder dressed in retroreflective clothing.[17] There were statistically significant advantages for both identifying the pedestrian and for recognition distance. Only 5 percent of drivers detected a black-clad pedestrian when the driver used low beams and faced glare, but 94 percent of drivers detected pedestrians wearing retroreflective markings that showed motion under favorable lighting conditions. Another study concurred that retroreflective trim on the arms of nighttime work zone workers was most helpful for visibility because it showed arm motion.[18]

Retroreflective clothing depends largely on reflecting headlights and is not as visible outside the headlight beams. However, electroluminescent materials were found to significantly improve the visibility of pedestrians beyond retroreflective apparel.[19]

While advice is sometimes given to pedestrians to take a **flashlight** after dark, no research was identified on the safety benefits. A flashlight can provide some illumination of potential walkway hazards, but it has limited practical value for increasing the visibility of the pedestrian or illuminating hostile individuals.

Retroreflective pavement markings and raised pavement markers can amplify the effectiveness of vehicle headlamps in helping the driver see lane and crosswalk lines and pavement legends. Raised pavement markers are useful to keep drivers safe in their lanes, especially in poor visibility conditions, such as wet weather. One analysis of LED-raised pavement markers noted that these could be used to "improve the safety of intersection approaches, as well as pedestrian, bicycle and other crossings."[20] However, no research was found directly quantifying the benefits for pedestrians.

Notes

1. Naomi J. Miller, Terry K. McGowan, and Rita N. Koltai, *Pedestrian Friendly Outdoor Lighting* (Richland, Washington: Pacific Northwest National Laboratory, 2013), https://www.semanticscholar.org/paper/Pedestrian-Friendly-Outdoor-Lighting-Miller-Koltai/f463f1476d0733506d6b1e6a3a188e8923a00ea1.
2. Ronald Gibbons, "Lighting and Pedestrians," Presentation to the 2020 IES Street and Area Lighting Conference, Dallas, Texas, October 26, 2020.
3. Eric Baldwin, "An Architect's Guide To: Outdoor Lighting," Architizer website, December 5, 2018, https://architizer.com/blog/product-guides/product-guide/outdoor-lighting/.
4. National Academies of Sciences, Engineering, and Medicine, *Solid-State Roadway Lighting Design Guide: Volume 2: Research Overview* (Washington, DC: The National Academies Press, 2020), doi: 10.17226/25679.
5. Los Angeles Bureau of Street Lighting, "Changing our Glow for Efficiency," Municipal Solid State Lighting Consortium – LED Workshop, Los Angeles, April 2012, http://docplayer.net/18037759-City-of-los-angeles-changing-our-glow-for-efficiency-municipal-solid-state-lighting-consortium-led-workshop-los-angeles-april-2012.html.
6. Steve Taggart, "City of Chicago Smart Lighting Program: From Idea to Realization," Street and Area Lighting Conference, Illuminating Engineering Society, October 27, 2020.

7. Bruce Kinzey and Jason Tuenge, "Show Me the Data: Does LED Lighting Influence Roadway Safety?" webinar, Illuminating Engineering Society, July 2, 2020.
8. National Academies of Sciences, *Solid-State Roadway Lighting Design Guide*.
9. National Academies of Sciences, *Solid-State Roadway Lighting Design Guide*, 3.
10. National Academies of Sciences, *Solid-State Roadway Lighting Design Guide*, 19.
11. "Outdoor Lighting Strategy," City of Vancouver website, accessed March 19, 2021, https://vancouver.ca/streets-transportation/outdoor-lighting-strategy.aspx.
12. City of Vancouver, Planning, Urban Design and Sustainability Department and Engineering Department, "Administrative Report: Outdoor Lighting Strategy," June 25, 2019, https://council.vancouver.ca/20190723/documents/a4.pdf; Erik Bonderud, City of Vancouver Engineering Services, Email, January 25, 2021.
13. National Academies of Sciences, *Solid-State Roadway Lighting Design Guide*, 22.
14. Paul Lutkevich, "Roadway Lighting Design," *Illuminating Engineering Society webinar*, September 8, 2020.
15. Christopher J. Edwards and Ronald B. Gibbons, "Relationship of Vertical Illuminance to Pedestrian Visibility in Crosswalks," *Transportation Research Record* 2056, no. 1 (January 1, 2008): 9–16, doi: 10.3141/2056-02.
16. San Francisco Municipal Transportation Agency and University of California Traffic Safety Center (Now SafeTREC), *Pedestrian Safety Engineering and Intelligent Transportation System-Based Countermeasure Program for Reduced Pedestrian Fatalities, Injuries, Conflicts, and Other Surrogate Safety Measures* (Washington, DC: FHWA, 2013), https://safety.fhwa.dot.gov/ped_bike/tools_solve/ped_scdproj/sf/.
17. Richard A. Tyrell et al., "On-Road Measures of the Visibility of Pedestrians at Night," *Journal of Vision* 3, no. 9 (October 2003): 549, https://jov.arvojournals.org/article.aspx?articleid=2129625.
18. James R. Sayer and Mary Lynn Mefford, "High Visibility Safety Apparel and Nighttime Conspicuity of Pedestrians in Work Zones," *Journal of Safety Research* 35, no. 5 (February 2004): 537–546, https://www.researchgate.net/publication/8189765_High_Visibility_Safety_Apparel_and_Nighttime_Conspicuity_of_Pedestrians_in_Work_Zones.
19. Drea K. Fekety et al., "Electroluminescent Materials Can Further Enhance the Nighttime Conspicuity of Pedestrians Wearing Retroreflective Materials," *Human Factors* 58, no. 7 (2016): 976–985.
20. Carlos Ibarra and Ed Rice, "FHWA Summary: LED Pavement Markers," FHWA Safety website, May 2009, https://safety.fhwa.dot.gov/intersection/conventional/unsignalized/tech_sum/fhwasa09007/fhwasa09007.pdf.

Chapter 6

Innovative Technologies

6.1 Purpose and Scope of This Chapter

The previous chapter addressed standard lighting equipment choices. This chapter discusses innovative technologies that are either immediately available, but not yet widely adopted, or expected within the next few years.

   This chapter first discusses "smart lighting" and the role of lighting in the Internet of Things (IoT). This first section covers dynamic illumination options, networked lighting, and the use of lighting equipment for data collection and municipal services for purposes beyond illumination. Next, this chapter covers non-standard or experimental mountings (bollards, in-pavement, and linear lighting). Finally, it discusses solar lighting.

   Chapter 9 discusses placemaking and aesthetic options, including some that take advantage of these new smart lighting technologies, such as programmed LED lighting shows. Chapter 10 covers potential engineering advances in the more distant future.

6.2 "Smart Lighting" and "Smart City" Applications

LED lighting has opened a range of possibilities for improving illumination and creating a smarter, networked city. LED-enabled "smart lighting" technology was listed as one of the ten most important recent urban innovations by the influential World Economic Forum.[1] However, technology is changing so rapidly that there is limited experience with many applications.

6.2.1 Overview of Features and Options

Smart lighting can be divided into three broad categories. Examples of each include:

1. Dimming and adjusting light levels ("adaptive lighting")
   - Light levels programed by the time of day
   - Dynamically adapting to the presence of pedestrians or other conditions

   DOI: 10.4324/9781003149750-6

- Increasing light output to offset lumen depreciation throughout the life of the luminaire
- Adjusting color correlated temperature (CCT) or other spectral features
2. Remote control and maintenance of illumination ("connected lighting")
   - Adjusting illumination centrally for a special event or emergency needs
   - Automatic notification of failures or other maintenance needs
   - Luminaire temperature detection
3. Monitoring, sensing, and electrical service mounted on light poles ("smart poles" with "smart city" applications)
   - Direct Short Range Communication with "connected vehicles"
   - Public Safety – monitor license plates to trace wanted vehicles and to be used as criminal investigation evidence or use gunshot/explosion detectors to alert first responders
   - Air Quality – Analyze the link between traffic conditions and pollutant levels, possibly suggesting locations for additional control measures
   - Traffic Optimization – Detect speeding and analyze volumes and patterns
   - WiFi, Bluetooth, and Cellular (4G and 5G) transmitters
   - Pedestrian Analytics – measure changes in pedestrian volumes; monitor pedestrian movements outside of a crosswalk
   - Road Condition Monitoring – Optimize salting/sanding routes by measuring road temperature
   - Flood Sensing – For emergency response

Technology has become far more flexible and powerful, but also more complicated with these options, which may be implemented simultaneously or staged over multiple projects. Also, lighting engineers and designers are now confronted by a range of technical and policy concerns that go well beyond illumination.

Investments in smart lighting could total $8.2 billion over the next decade, according to a survey from smart infrastructure market intelligence firm Northeast Group LLC.[2] Navigant Technologies forecast a 28.5 percent compound annual growth rate in the smart street light market through 2028.[3] A survey found 19 percent of large U.S. cities considering smart lighting, including remote street light monitoring and using street lights to support the Internet of Things. Smart lighting projects have been pursued by metropolitan cities, such as New York, as well as very small cities such as Spencer, Massachusetts, with a population of 12,000.

As LED street lighting has become widespread, LED features and flexibility have facilitated a dramatic increase in "smart lighting" capabilities: finely adjusting light levels, networking lighting for remote control and maintenance, and housing sensors and electrical services on light poles for non-illumination applications.

6.2.2 Benefits and Potential Adverse Impacts

Adaptive lighting's benefits arise chiefly from greater control and flexibility of lighting. LEDs allow control of light output, color correlated temperature, chromaticity, and spectral power distribution.[4] Adaptive lighting has several potential advantages by providing "just right" illumination levels (in the phrase of Ronald Gibbons of Virginia Tech Transportation Institute).[5] Dynamic lighting can help decrease lifecycle costs directly through reduced energy and maintenance needs by reducing illumination levels during periods when less light is needed. Reducing light levels should also reduce light pollution. Increasing illumination when a pedestrian is detected entering a crosswalk could provide additional warning to drivers while improving the visibility of pedestrians. Adjusting the color appearance and frequency distribution of lighting also has potential visibility and pedestrian reassurance benefits.[6] It could possibly even minimize any adverse melatonin suppression impacts (although such impacts are likely very small, as discussed in Chapter 4).

Any reductions in light levels through adaptive lighting need to minimize adverse impacts on safety. Virginia Tech Transportation Institute (VTTI) researchers compared illumination levels (using several lighting metrics) with night-to-day crash rate ratios on different street, highway, and freeway segments.[7] Observations generally supported Illuminating Engineering Society (IES) recommended levels for most streets and highways but suggested that light levels could be reduced "during periods of reduced traffic and potential conflict" while still meeting IES criteria. (For example, during late-night periods of low pedestrian volumes, illuminance on a major street with high daytime pedestrian volumes could be dimmed to correspond to the major street, low pedestrian volume classification.)

Connected lighting benefits derive from giving central stations greater information and control over lighting remotely to improve the quality of lighting and reduce costs, including real-time adjustment capabilities. Taking advantage of such capabilities requires adjusting departmental procedures. For example, maintenance calls can be dispatched quicker and more efficiently, but the information on lighting failures must be routed correctly. Systematic problems and inventory needs can be identified by analysis of data on lighting performance, but lighting managers need to use this information. With the right equipment, light output at specific locations can more easily be increased for special events or local emergencies, but that requires coordination with public safety and event planners.

Smart poles can host sensors for remote monitoring of non-lighting data to benefit traffic safety, public health, and other civic needs. These capabilities are not generally new, but the poles provide nearly ubiquitous staging locations with electricity connected to a central station. Smart poles can also be used

Adapting light levels dynamically to the minimum needed for safety has potential benefits of reduced energy use, costs, and light pollution.

for electric vehicles and device charging/broadcasting. Capabilities will likely evolve with technology changes. Eventually, Direct Short Range Communication (DSRC) between connected vehicles and infrastructure could, for example, warn drivers or control their vehicles to avoid hazards such as a red-light runner, obstacle in the roadway, unexpected queue, or flooding.

The **potential downsides** of all smart lighting start with increased costs. Besides the additional installation costs, which can run in the millions (or even tens of millions) of dollars for a large city, wireless communications devices and new sensors carry their own maintenance needs, which may be hard to predict. However, adaptive lighting promises a potential payback for initial capital costs through reduced electricity and maintenance costs. The ability to pay for additional features through energy savings depends in part on technical standards for adaptive lighting and on utility tariffs.

Potential privacy intrusions and data ownership questions have already triggered concerns by residents. For example, watchdog group San Diegans for Open Government (SDOG) sued the City of San Diego over not releasing data collected through its Smart Street Lights Program, whose sensors gathered a wide range of information, including pedestrian and traffic movements, and also recorded video that police say they have used to solve violent crimes.[8] Video surveillance by the Chinese government of individuals using facial recognition understandably makes many U.S. citizens concerned about the potential for similar abuses.

Automated communication with in-vehicle devices raises concerns about security. For example, will such features possibly increase the risk of hacking vehicle electronic systems?

Aesthetic and health issues may also arise. The Los Angeles City Bureau of Street Lighting recognized the potential for street light poles to look like "hat-racks" with multiple sensors and transmitters. Residents are often opposed to cell phone micro-transmitter stations based on aesthetic and health concerns.

6.2.3 Process for Designing and Implementing Smart Lighting

Procedures for planning and implementing smart lighting projects were suggested by the International Nighttime Design Initiative (NTD), based partly on a pilot study for Saratoga Springs, New York, in cooperation with government agencies in the Capital District of New York State.[9] This guide is intended to serve communities with a range of interests, from implementing a simple LED conversion to a more ambitious program of smart city applications.

The first step is to **determine what problems or challenges** smart lighting can address. A range of stakeholders and technical experts are needed to determine this, including many who are not lighting specialists. For example, public sector representatives often include public works, transportation agencies, law enforcement, and information technology. Multiple levels of government, community representatives, utilities, manufacturers, consultants, and contractors are typically involved.

The second step is to develop an **overall strategy**. It needs to address financial and technological constraints and benefits realistically. For example, electricity savings are often an important source to justify and fund capital costs, but monetizing these savings may be complicated if electricity costs are governed, not by actual electricity usage, but by tariffs (agreements with utilities on cost reimbursement). Maintenance savings also can contribute to a positive return on investment, but they are difficult to forecast.

Often it is best to start with one or a very few **pilot programs.** By achieving quicker accomplishments, smart lighting initiatives can build relationships and develop confidence among stakeholders.

Design needs to account for rapid change in technology in part by building in flexibility. The ability of devices (often manufactured by different companies) to work together (interoperability) is a key concern. Accessibility of devices for maintenance, aesthetics, and device communications is also an important consideration.

Approvals for smart lighting are often far more complicated than for conventional lighting installations. For example, the contracts approved by local government policy bodies may deal with highly innovative technologies and complicated maintenance provisions.

Procurement of the system can be accomplished in stages, or the locality may seek a turnkey system in which one contractor or team is responsible for the entire project. Expertise in IT is often essential.

Installation may require specialized contractors with expertise in IT and communications. Testing of equipment and training government staff to take responsibility are important and potentially complicated by diverse technical standards.

Local government needs to manage **operations and maintenance** actively. This involves both technical oversight and interaction with stakeholders to respond to concerns and requests. There may be numerous agency and citizen requests for data provided by sensors, possibly with special processing.

Initial efforts or pilots should be **evaluated**. The program can be adjusted, and additional projects planned as needed and feasible.

6.2.4 Adaptive Lighting: Options and Considerations

Adaptive lighting may use time-based dimming or motion detectors to trigger different illumination levels. Time-based dimming is not recommended by the IES *Recommended Practice* RP-8–18 for:

- Signalized intersections
- Midblock crosswalks
- Roundabouts
- Railway crossings

However, raising illumination *above* recommended levels when a pedestrian is detected may be considered consistent with the intent of the guidance.

The pedestrian is often key for adaptive lighting: Light levels may be adjusted by a pre-programed schedule based on expected pedestrian volumes or adjusted dynamically by motion detection of pedestrians. Advances in LED lighting features allow greater control of CCT, glare, and light distribution, as well as improved maintenance and monitoring.

Different strategies may be used or combined:

- Initial light output may be generally lowered to "maintained levels" (i.e., illumination levels are typically set higher than standards but decrease gradually over time closer to the recommended levels as luminaires get dirty or performance diminishes)
- Specific locations may be dimmed if found to have higher levels than necessary
- Illumination may be set to correspond to pedestrian levels (predicted or measured in real time).

Pedestrian activity is a key factor in determining illumination levels, along with street classification. Pedestrian activity typically is based on a count or an estimate of pedestrians on the sidewalk on a single block for a one-hour period of relatively high after-dark activity. Lighting designers generally have minimal pedestrian volume data. While pedestrian activity for most residential and industrial locations can be roughly estimated, pedestrian volumes for commercial and special event districts can be highly variable and difficult to predict. Thus, pedestrian detection can be especially useful in these cases.

A detailed procedure to setting specific light levels is suggested in the FHWA *Adaptive Lighting Guide*.[10] It recommends five illuminance levels for a location that depend on a weighted combination of the following factors:

- Vehicle speeds
- Traffic volumes
- Median presence
- Intersection density
- Ambient luminance
- Signage and pavement markings
- Pedestrian volumes
- Parked vehicles
- Possible need for facial recognition

Adaptive lighting technologies are sensitive and complicated. Motion detectors, for example, may be unintentionally triggered by animals, blowing leaves, or wind.

Advanced LED features with promise for widespread use in outdoor lighting within the next several years include tunable white light LEDs (with control of spectral content), greater control over light distribution, low-glare luminaires, and programmable lighting. Also, these lights could be adjusted for warmer (higher) CCTs later at night to address concerns about possible melatonin suppression. Lights could also be adjusted remotely to cooler, brighter CCT when an emergency is detected in the area. (White Bear Lake, Minnesota, has installed such lights in its central Railroad Park.[11])

6.2.5 Adaptive Lighting: Field Tests and Permanent Installations

Several field tests of dimming or adjusting light levels have been conducted. These include cases with both preprogrammed illumination levels and pedestrian motion detection.

The City of Seattle installed adaptive lighting as standard in its Belltown district. Illumination levels were set to vary according to traffic and security needs: 80 percent from dusk until 1 AM, then 100 percent when nightclubs let out from 1 AM to 5 AM. Seattle considered additional installations that could adjust light levels via pedestrian activation or according to traffic levels.

Cambridge, Massachusetts, included an adaptive control system when it replaced its 7,000 street lights and specialty/park fixtures with LEDs. A wireless control system enables operators to dim street lights to 70 percent of initial light levels. Later in the evening, lights dim further to about 35 percent of their initial brightness, consistent with lower pedestrian levels. "Based on our experience in Cambridge, I think that fears that lowered light levels will be perceived as unsafe are unfounded," reported Glenn Heinmiller, a design consultant involved in the project. "In our city of 100,000 residents, we've had no comments about lower levels. No one has complained or even seemed to notice when the light levels are cut in half."[12]

As part of a federally funded test of pedestrian safety countermeasures in three U.S. cities, University of Nevada, Las Vegas researchers found a statistically significant increase in driver yielding by using supplemental lighting for a mid-block crosswalk, triggered by automated detection of pedestrians approaching the uncontrolled crosswalk.[13] Pedestrian detection was provided by microwave and infrared devices. The devices have since been removed.[14]

The evaluation found a reduced level of jaywalking outside the crosswalk and reduced cases of pedestrians being "trapped" in the crosswalk. Possibly the illumination convinced more pedestrians to cross at the crosswalk and made their crossing easier by leading motorists to yield more. However, due in part to the relatively high cost of installation, this was not judged one of the most cost-effective measures when compared to a broad range of generally low-cost physical measures by the Federal Highway Administration (FHWA) and Las Vegas, and it was removed. The evaluation noted that "While driver yielding increased, its prevalence was still low at 35 percent." In Miami, the addition of dynamic lighting to a crosswalk that had Rectangular Rapid Flash Beacons (RRFBs) did not appear to further improve driver yielding or pedestrian-vehicle conflicts. The Miami researchers suggested that this may have occurred because the dynamic lighting is not very noticeable in the presence of highly intense flashing beacons. (More cost-effective measures were judged to be leading pedestrian intervals, pedestrian countdown signals, in-street pedestrian signs, activated flashing beacons, Rectangular Rapid Flash Beacons, and pedestrian push buttons that confirmed when a call was received. Another effective method was a Danish offset, forcing pedestrians to face oncoming traffic before leaving a median island, combined with high-visibility crosswalks, advance yield markings, and YIELD TO PEDESTRIAN signs.)

Superior vehicle driver detection distances for LED lighting were described in Chapter 5. Tests in Seattle and San Jose suggested that dimming LED luminaires to 50 percent levels did not significantly decrease object detection distance with dry pavement, and even dimming to 25 percent did not markedly reduce detection distance.[15] Based partly on this research, the City of San Jose adopted guidelines for adaptive lighting in its *Public Streetlight Design Guide*. These guidelines are described in the next chapter.

Los Angeles plans mid-block crosswalks that increase in illumination when pedestrian activity increases. It also has transition zones around sports venues that brighten during times of events.

Adaptive lighting: case study: University of California at Davis

The University of California (UC) at Davis cut outdoor lighting energy costs by 86 percent by using lighting on campus paths that adapts illumination levels to pedestrian activity.[16] Under the campus "Smart Lighting Initiative," over 1,500 network-controlled LED street lights, area lights, post-tops, and wall packs were installed in 2012. (See Figure 6.1.)

New 40 watt and 90 watt LED fixtures were installed for roadway and area lights. These reduced lighting energy use by 73 percent. With maintenance savings included, these retrofits were forecast to achieve payback in 12.9 years, despite low electricity rates.

New LED wall packs (building mounted lights) operated at 42 watts, automatically dimming from 80 to 90 percent of full output to 20 percent when no pedestrian occupancy was detected. These wall packs had energy savings of 89 percent and a payback of 9.9 years.

LED post-top luminaires consumed 45 watts with a typical occupancy rate of 40 percent. Average energy savings were 87 percent, with a payback of 12.8 years.

This effort responded to California's energy efficiency codes and standards for new and altered outdoor lighting for nonresidential and high-rise residential buildings (2019 Title 24, Part 6), requiring generally that luminaires that consume 40 watts or more and are mounted at a height of 24 feet or less must be controlled with motion sensors or automatic scheduling dimming.

▲ Figure 6.1

Networked Adaptive LED Lights at the University of California, Davis. Courtesy of California Lighting Technology Center - University of California, Davis.

6.2.6 Connected Lighting: Options and Considerations

Connected or "networked" lighting uses centralized computer stations and wireless or wired links to control illumination, detect maintenance needs, and measure the power usage of street lights in a district or city. Typically, a data logger collects and stores information on the status and operation of the electrical and lighting system. Each street light may have its own control, measurement, logging, and communications features. (See Figure 6.2.) In addition to real-time control, connected lighting systems enable a range of data analyses, such as preventive maintenance, equipment inventory and procurement, workflow management, and measurement of electricity usage.

Numerous issues must be considered in designing the network per the NCHRP publication *Solid-State Roadway Lighting Design Guide:*[17] These include:

- Adaptive control technology
- Equipment cost, reliability, and performance
- Network structure issues (e.g., signal reliability, acceptable distances between wireless devices)
- Data communications type (wired, RF, cellular) and accuracy

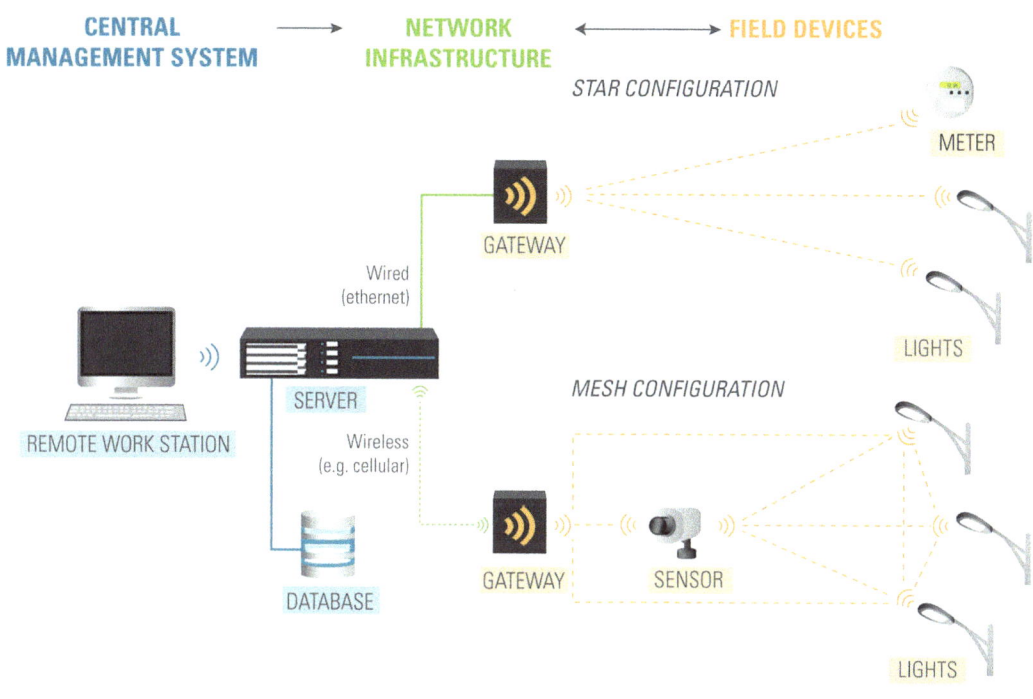

CENTRAL MANAGEMENT SYSTEM → NETWORK INFRASTRUCTURE ↔ FIELD DEVICES

STAR CONFIGURATION

METER

GATEWAY

Wired (ethernet)

REMOTE WORK STATION

SERVER

Wireless (e.g. cellular)

LIGHTS

MESH CONFIGURATION

DATABASE

GATEWAY   SENSOR

LIGHTS

▲ Figure 6.2

Typical Components and Connections for Networked Lighting System. Courtesy of California Lighting Technology Center - University of California, Davis.

- Cybersecurity and privacy
- Changes to current policies and procedures (e.g., through automated maintenance work orders, changes in asset management)
- Method of measuring power usage
- Integration with new non-lighting Smart City applications
- Interoperability (the ability of devices, apps, and systems to reliably and securely exchange data)
- Flexibility for future changes
- Staffing and consulting needs

Obviously, the skills needed to design and operate networked lighting go beyond traditional lighting engineering, requiring expert IT and communications technology support. For example, Geographic Information Systems (GIS) are typically used as a map-based database for equipment. The systems may

Networked lighting, possibly with non-illumination Smart City applications, is far more complicated than conventional street lighting. Significant Information

Technology expertise is required, and major changes to municipal work practices may be needed.

include sophisticated features like integration with the Computer-Aided Design (CAD) system and asset management software.

Cybersecurity and privacy protections are the responsibility of all stakeholders.[18] Manufacturers and contractors should address these issues through project analysis and product design, in training personnel, and providing incident support. Knowledge of regulations and standards is critical. Customers need to be sure they understand how to securely use equipment and they are keeping up with device updates.

Advances can be expected in the next several years in the range of connections, the degree of automation facilitated by networked lighting, and the reliability of systems. Smarter use of data should provide improved luminaire failure detection and prevention, improved maintenance workflow, and the use of motion-sensor data for transportation planning and other purposes.

6.2.7 Connected Lighting: Field Tests

Networked lighting has been installed by some of the largest cities in the world, including Los Angeles, San Diego, Chicago, Barcelona, Copenhagen, and London. In implementing these projects, these cities aimed for energy savings, improved maintenance procedures, and economic development. While these projects are promising, it is too early to evaluate whether goals will be achieved.

Chicago executed a $150 million contract in 2017 for LED conversion of some 270,000 lights, installation of a lighting management system, equipment repair, integration of 311 problem reporting and equipment GIS, and electricity monitoring for rebate support.[19] The process included a Request for Information (RFI), followed by a Request for Qualification (RFQ), Request for Proposals (RFP), and negotiations culminating in a Best and Final Offer. For decorative luminaires, the project removed the fixtures and installed the LED lamp and devices in a local assembly facility, using workers from disadvantaged neighborhoods. Extensive time was needed to check that photometrics met IES RP-8-18 guidance.

In 2017, San Diego initiated a $30 million program for the "world's largest smart city IoT sensor platform, which was suspended in 2020 over privacy issues and funding (as described further below)."[20] (Sensors use the poles for mounting height and power, but they have a separate communication system from the illumination cellular network as discussed in another section below.) The project includes 14,000 new LED fixtures, networked with wireless technology. Mesh-networking technology allows centralized dashboard switching, dimming, brightening, and maintenance checks. Energy usage is also metered in the luminaires, with the potential for additional savings beyond just the LED conversion.

The Los Angeles Bureau of Street Lighting has a GIS database for over 200,000 street lights, with data on lighting attributes and electronic links to engineering plans for specific installations.[21] (See Figure 4.1.) The database is used routinely for managing maintenance and repair, billing, and planning and design work. For example, failures can be automatically reported, and a GIS user can query the equipment at a particular location. Incident or damage reports are stored, including images and links to repair jobs. Crew and equipment needs for specific repairs can be determined and tracked. Such information also allows offline analyses of the productivity and effectiveness of personnel and equipment.

6.2.8 Smart Poles: Options and Considerations

Street light poles are excellent tools for housing new sensors and devices like electric vehicle charging because they are powered, ubiquitous, and at a useful height. LED conversions also allow cities to add new non-lighting monitoring technologies and smart city apps in a more cost-effective manner through combined construction projects. However, sensors and other devices can also be added separately. The communications network to support these non-lighting capabilities also can be integrated with the network for remote control of lighting, but it often needs to be separate. For example, the video capabilities in the San Diego network require a higher-speed wireless technology than provided by the LightGrid illumination network, such as 4G or eventually, emerging 5G systems.

The potential **applications** listed earlier in this chapter span transportation, public health, public works, and public safety uses. Data from sensors can be used in real time (e.g., for traffic control) or analyzed offline.

Because non-illumination sensors and cameras on light poles involve very new technology, separate from lighting technology, this topic is not covered comprehensively by standard lighting **technical references**. The International Nighttime Design Initiative *Municipal Smart City Street Light Conversion and Evolving Technology Guidebook* addresses "smart city" applications as part of its smart lighting treatment. The Los Angeles Bureau of Street Lighting's *Strategic Plan* discusses its smart poles in detail.[22] A number of websites also offer more up-to-date information.[23]

Highly innovative approaches require a complete **shift in planning, design, and operation** of street lighting equipment from the traditional. For example, in planning for smart city applications, a range of concerns about privacy, cybersecurity, operations, and maintenance need to be addressed. Financial and technical considerations can be highly complex, and the stakeholders are potentially far more varied than traditionally consulted. Advanced technical capabilities in the vendor team and overseeing government staff are critical.

Clients for data and services include government agencies, citizens, and businesses. The level of access and sharing needs of such a diverse group of clients should be determined in advance. Electric vehicle charging stations or WiFi transmitters need to be planned in coordination with large non-lighting projects and entities.

Street light poles are prized as potential hosts for smart city sensors and other equipment because of their ubiquity, power, and height

Smart poles are typically heavier and more expensive than conventional poles in order to support multiple sensors, cameras, and other devices.[24] Aesthetic concerns about a possible undesirable "hat-rack" appearance may need to be addressed. The spacing and siting of poles may shift to enable specific applications like video monitoring of an intersection.

Operation of "smart city" applications can involve extensive interface with stakeholder and citizen requests for data. For example, small businesses may request parking usage data for their blocks. Citizen advocacy groups may expect to review any police-related video footage due to privacy concerns.

6.2.9 Smart Poles: Field Tests

A range of cities, large and small, have invested in "Smart City" applications housed on smart poles. Installations are generally too recent to provide a comprehensive evaluation.

San Diego started deploying the largest smart city lighting-based project undertaken by a U.S. municipality (but as of February 2021 had frozen implementation).[25] At that point, San Diego had installed about 3,000 of the planned 4,300 intelligent sensor ("CityIQ") nodes (controllers), in addition to the 14,000 new LED fixtures. The CityIQ nodes are being installed citywide, on selected street light poles, both conventional and historic (in the Gaslamp district). Nodes are installed as densely as every 100 feet downtown. The cost of installing the smart sensors was approximately $11 million.[26]

Cameras and other sensors, connected with high-speed cellular technology, collected generally anonymous data. The City envisioned not only municipal applications, but also sharing its data with organizations and businesses. Sensors were planned to detect air quality, temperature, wind speed, and other data. Forward-and rear-facing video cameras (providing a 360-degree field of view) were also installed. Transportation managers had data visualization apps for parking and traffic safety. Police used ShotSpotter technology to detect the location of gunshots. More than three years after the start of the project, data were mostly not available in a useful form to the public for pedestrian counts, parking occupancy data, or other applications, which City staff blamed partly on a cumbersome contract with a vendor.

The ambitious project was put on hold due to citizen opposition and funding issues.[27] A watchdog group, San Diegans for Open Government, sued the City for allegedly not responding to its request to release 24 hours of raw data and all records of processed data.[28] Activists argued for a surveillance ordinance and a privacy advisory commission, which both later were supported by the City Council, and the Mayor ordered the cameras to be disabled.[29] (Seattle also has a surveillance ordinance that addresses policies for the acquisition and use of surveillance technology, aiming for greater public transparency and equity.[30])

The video cameras have been especially controversial, although the vendor stated that footage has been used to help the San Diego Police Department and the San Diego court system solve more than 350 violent crimes over a two-year period.[31] The video capability could also provide precise data collection for parking occupancy, pedestrian counts, potholes, landscape, and trash issues. However, video has also made the installation process more complex and controversial even for non-police applications. For example, the cameras are set up electronically to avoid collecting data on private property, which requires mapping precisely pixel by pixel what is public versus private property.

Funding has also been insufficient to keep the project active. Energy savings from the LED installation are intended to help fund the project, but the major initial costs are a challenge. The use of federal Community Development Block Grant funds was also questioned by City Council members.[32]

Barcelona installed 39,000 LED street lights, about 18,000 with pole-mounted sensors for monitoring air quality, noise, electricity usage, pedestrian and vehicle movements, plus Wi-Fi transmitters.[33] Sensor data are shared on the internet, with open-source software available via Github to allow urban planners and others worldwide to study the data.[34]

Syracuse, New York made upgrades and added controls to monitor and manage LED lighting.[35] The "Syracuse Surge" smart city plan includes flood sensing, air quality and road temperature monitoring, and vacant structure monitoring, which also includes illegal dumping on these sites. Traffic counting may be added. Spencer, Massachusetts initiated a road temperature monitoring pilot project, using sensors installed as part of its smart lighting project.[36] The sensors should help the town to target road salting to sections most likely to freeze, with potential safety and financial benefits.

Rockhampton, Australia installed a post-top system. This included integrated speakers with ambient music and civic disaster messaging and duress buttons to initiate security intervention.

Los Angeles has started widespread installation of smart poles that would house a range of devices within the pole[37] (See Figure 6.3.) The Los Angeles smart poles and conventional poles house a range of monitoring and services, including:

- 432 electric vehicle charging stations
- Air quality meters
- Gas company communications equipment for smart meters
- 852 solar panels to generate electricity not just for the street lights but for the electric grid
- Cameras for surveillance, and video analytics to count cars, pedestrians, and bicyclists
- Speakers
- Electrical outlets

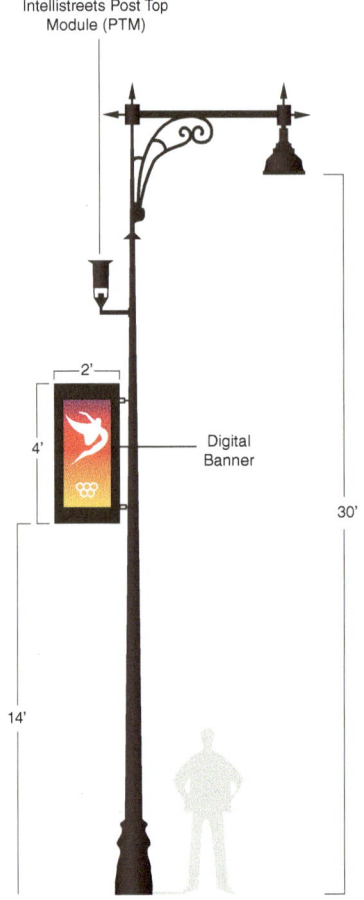

Intellistreets Post Top
Module (PTM)

2'

4'

Digital
Banner

30'

14'

◀ Figure 6.3

Los Angeles "Smart Pole". Courtesy of City of Los Angeles Bureau of Street Lighting.

Smart Pole Features
May include:

- 5G
- Air Quality Sensors
- Digital Banner
- Multi-color Lighting
- Pedestrian & Mobility Counter
- USB
- Wi-Fi

- USB outlets
- Wi-Fi capability
- Over 2,500 4G and 5G attachments for residents, as well as for first responders during an emergency situation.

Los Angeles developed an equity and outreach strategy for its smart lighting pilots to cover a range of concerns.[38] The outreach effort used the concept of "digital inclusion." The City also selected a street light pole design template through an open contest. The "Superbloom" design includes an LED strip for announcing events and emergencies, plus options for a shade umbrella and smart city attachments.[39]

Municipal and private utilities are seeking new revenues from smart city uses of their poles. For example, the Tampa, Florida, Police Department obtained grant funding for a five-year pilot on gunshot monitoring sensors mounted on utility street light poles.[40] The utility executed Smart City Public Safety Agreements to govern the provision of pole space and power.

6.3 Innovative Mountings

The type of mounting can substantially change light distribution, contrast, and other factors that influence pedestrian safety and walkability. Innovative mountings may also provide greater flexibility in installing lights in challenging environments, such as narrow rights-of-way.

6.3.1 Bollards

Lights on bollards (posts typically under four feet tall) are common for off-street outdoor lighting (like pathways, plazas, or wide stairways), but they are rarely used for crosswalks or sidewalks. (See Figure 6.4.) They are less intrusive visually than conventional street lighting poles or typical post-top pedestrian-scale lighting, although they can be a physical obstruction to pedestrians.

Due to their low height, bollards provide a relatively limited light distribution. They are also especially susceptible to vandalism but are easier to maintain than taller poles.[41] Bollards may not provide sufficient light for facial recognition. They may be affected by snow clearance needs.

IES *Recommended Practice* RP-8-18 suggests that "bollards are not recommended for use on or adjacent to roadways, because their low mounting heights could be distracting to road users."[42] That relates to a concern about potential glare from luminaires directly in the field of vision of drivers.

▶ Figure 6.4

Typical Bollard Lighting. Courtesy of Adobe Stock Photos. © London Time - stock.adobe. com.

While bollards are routinely used for lighting walkways and plazas, crosswalk use is rare. Research has suggested advantages of high contrast for pedestrians and low glare, but there are concerns about limited facial recognition by pedestrians and possible distraction for drivers from the luminaires themselves.

There have been limited field tests of their use for crosswalks. Researchers at Rensselaer Polytechnic Institute's Lighting Research Center found some "promising" advantages for bollards over overhead lighting in tests in Aspen, Colorado, Schenectady, and Slingerlands, New York, and Old Bridge, New Jersey.[43] Researchers designed lighting to provide at least 10 lux vertical illumination at three feet above ground within the crosswalk, while limiting vertical illuminance at five feet above ground to a maximum of one lux, thus reducing glare. They used full-spectrum "white light" to enhance brightness and contrast with existing overhead high-pressure sodium (HPS) illumination. (The test bollard lighting is not used in Aspen today.[44])

Bollards illuminated pedestrians more than the background, providing more contrast and better relative visual performance (RVP) than overhead HPS. They maintained a "positive" contrast, without interference from vehicle headlights. Pedestrians in Schenectady rated the bollard lighting positively, and most pedestrians felt "somewhat" or "completely" safe and secure in the crosswalk. They did not report glare or dark areas.

6.3.2 Linear Overhead Lighting

Linear overhead lighting can be used when overhead structures cover walkways or pedestrian nodes like transit centers. For example, the underside of a bridge may be used for lighting a trail or a long transit shelter may light a bus boarding area.

The primary benefits of such lighting are in some cases the convenience and relatively low cost of mounting since an existing structure is used in place of a pole. The overhead element may be able to carry electrical service more easily than it could be provided to a pole. There may be limited right-of-way or drainage issues that could constrain possible pole locations.

Long linear LED luminaires have been used in limited field tests of this technology. For example, Seattle has used linear overhead lighting under the Ballard Bridge. Seattle and Whistler, British Columbia, have also used this technology for bus shelters.

For its Ballard Bridge project, Seattle needed to provide a better pedestrian/bicycle connection between the Burke-Gilman Trail and RapidRide (bus rapid transit). This important link had an unpaved segment with poor lighting. There were a number of constraints that made it hard to provide adequate lighting: an extremely narrow (10–12-foot, 3–4 meter) right-of-way, utility constraints, limited electrical power sources, drainage issues, and funding limitations. As the linear lighting had relatively low power needs, it was possible to tap a solar energy source no longer being used. The solution was to install linear LED lighting on the underside of the bridge structure. (See Figures 6.5 and 6.6.)

Lighting was designed to achieve average illuminance of 1.0 footcandle (fc) with a uniformity ratio of 4.0 or less. Measured illuminance was better than this: 1.11 fc with a uniformity ratio of 2.78.

A Whistler, British Columbia, bus shelter at a transit hub also used linear LED luminaires in 78 triangular wood cells, notched to prevent light trespass and glare.[45] A photocell activates the shelter lighting when ambient light levels

Linear Lighting on Underside of Ballard Bridge Structure. Courtesy of Seattle Department of Transportation and James Le.

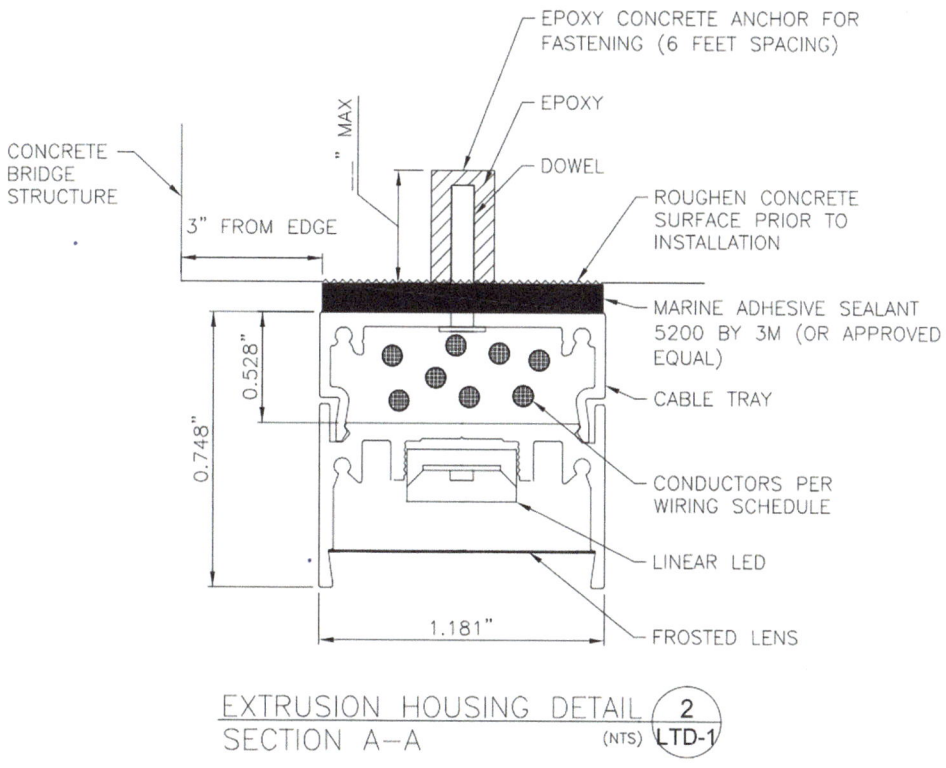

EPOXY CONCRETE ANCHOR FOR FASTENING (6 FEET SPACING)

EPOXY

CONCRETE BRIDGE STRUCTURE

" MAX

3" FROM EDGE

DOWEL

ROUGHEN CONCRETE SURFACE PRIOR TO INSTALLATION

MARINE ADHESIVE SEALANT 5200 BY 3M (OR APPROVED EQUAL)

0.528"

CABLE TRAY

CONDUCTORS PER WIRING SCHEDULE

0.748"

LINEAR LED

1.181"

FROSTED LENS

EXTRUSION HOUSING DETAIL ②
SECTION A—A   (NTS) LTD-1

▲ Figure 6.6

Design for Linear Lighting Housing. Courtesy of Seattle Department of Transportation and James Le.

are 1 fc or less. By recessing the lighting into the structure cells, the luminaires were invisible and vandal-resistant, but attractive features of the wooden frame could be highlighted.

6.3.3 In-Pavement Lighting

In-pavement lights include both in-roadway lights that delineate shoulder edge lines or crosswalks and walkway lights typically placed among pavers. In both cases, the objective is primarily guidance rather than full illumination of the roadway or walkway. Solar-powered lights may be installed.

In addition, flashing in-pavement crosswalk lights activated by pedestrians have been used as a traffic control device (as a warning to drivers), but these are less often installed than rectangular rapid flash beacons (RRFBs) and pedestrian hybrid beacons (or "Hawk signals"). RRFBs and pedestrian hybrid beacons provide only limited illumination of pedestrians but use eye-catching flashing patterns and warning signs to alert drivers.

For walkways, lights can be installed in between cobblestone or other pavers, either as the size of an individual paver or as a long strip. No research was found to evaluate the effectiveness of in-walkway lighting, but anecdotal evidence raises concerns about maintenance.[46]

6.4 Solar Lighting

Solar lighting has been used to only a limited extent by U.S. municipal governments, primarily restricted by battery performance limitations. However, by 2017 global sales of solar street lighting reached a cumulative 3.8 million units, with almost half of sales in the Asia/Pacific region.[47] They are forecast to exceed 9 million units worldwide by 2024.[48] Besides the obvious advantages of reduced energy use, solar lighting may be less susceptible to flooding or other interruptions to electrical service.

Los Angeles has used solar lighting on pedestrian/bicycle paths and has placed solar panels on its smart street light poles. Solar panels on street lights are intended not only to power the street light, but to feed back electricity into the grid. Las Vegas has also used solar lighting in a limited manner (on its Boulder Plaza, which also provides kinetic foot pads to help power the street lights).[49] Beaumont, California, a city of over 50,000, by policy only installs solar street lighting and has installed hundreds of poles.[50]

Lead-acid deep-cycle gel batteries are the primary means of household storage from solar panels, and are also useful for solar lights. Improved batteries are a promising area for new technology. Lithium batteries are increasingly used due to superior energy conversion efficiency, size, and longer service life.[51]

There are a number of other possibilities for improved solar cells. For example, Perovskite solar cells hold promise for a more efficient method of converting sunlight into electricity. They can also be lightweight, transparent, and printable. However, they are not commercially viable for widespread use, and their longevity needs to be demonstrated.[52] Cells that capture infrared light radiated from the ground at night are another option for study.

For completely new neighborhoods, a compelling alternative to individual solar fixtures is to collect solar energy with community solar gardens to power new lighting on a mini-grid.[53] Solar gardens centralize solar cell placement in convenient, efficient locations.

Notes

1. Ryan Holeywell, "Top 10 Urban Innovations to Pay Attention to," *Government Technology*, December 7, 2015, https://www.govtech.com/fs/perspectives/Top-10-Urban-Innovations-To-Pay-Attention-To.html.
2. Chris Teale, "Cities 'Finally Waking Up' to the Benefits of Smart Streetlights: Survey," Smart Cities Dive website, May 4, 2020, https://www.smartcitiesdive.com/news/survey-smart-streetlights-LED-northeast-group/577227/.
3. Rich Laezman, "Cities Getting Smart about Street Lighting," *Electrical Contractor Magazine*, January 2020, https://www.ecmag.com/section/lighting/cities-getting-smart-about-street-lighting.
4. Michael Poplawski, "Illuminating Thoughts on Nighttime Design: Technology Opportunities and More," National Association of City Transportation Officials Designing Cities Conference, Seattle, 2016, https://nacto.org/wp-content/uploads/2016/08/Nighttime-Design-Technology-Opportunities-and-More-Poplawski.pdf.
5. Ronald Gibbons, "Connected Infrastructure Activities," Pedestrian and Bicycle Information Center Webinar on "Lighting for Pedestrian Safety and Walkability," October 17, 2018, https://www.pedbikeinfo.org/webinars/webinar_details.cfm?id=13.
6. CIE (International Commission on Illumination), *Lighting for Pedestrians: A Summary of Empirical Data*, CIE 236 (Vienna: CIE, 2019), 8.
7. Ronald Gibbons et al. (Virginia Tech Transportation Institute), *Design Criteria for Adaptive Roadway Lighting*, Report FHWA-HRT-14-051 (McLean, VA: FHWA Office of Safety Research & Development, 2014), https://www.fhwa.dot.gov/publications/research/safety/14051/14051.pdf.
8. Teri Figueroa, "Government Watchdog Sues San Diego Over Smart Street Lights," *Governing*, December 19, 2019, https://www.governing.com/news/headlines/Government-Watchdog-Sues-San-Diego-Over-Smart-Street-Lights.html.
9. Plannng4Places and International Nighttime Design Initiative, *Municipal Smart City Streetlight Conversion and Evolving Technology Guidebook* (Albany, NY: Capital District Transportation Committee, 2020), https://nighttimedesign.org/connecting-communities-holistic-smart-lighting-enabling-technologies-guidebook-released/.
10. Gibbons et al., *Adaptive Roadway Lighting*.
11. Maury Wright, "Echelon Announces Tunable-White Outdoor LED Lighting Trial," *LEDs Magazine*, February 6, 2017, https://www.ledsmagazine.com/smart-lighting-iot/article/16700748/echelon-announces-tunablewhite-outdoor-led-lighting-trial.
12. Ronald Gibbons, Joseph Cheung, and Paul Lutkevich, "The Future of Roadway Lighting," *Public Roads* 79, no. 3 (November/December 2015), https://www.fhwa.dot.gov/publications/publicroads/15novdec/06.cfm.
13. "Hispanic Pedestrian & Bicycle Safety," US Department of Transportation, Federal Highway Administration, accessed March 19, 2021, https://safety.fhwa.dot.gov/ped_bike/tools_solve/ped_scdproj/sys_impact_rpt/chap_4.cfm#41.

14. Jerry Walker, City of Las Vegas Street Lighting, Operations and Maintenance, Email, December 5, 2020.

15. Nancy Clanton et al., *Seattle LED Adaptive Lighting Study – NEEA* (Portland, OR: Northwest Energy Efficiency Alliance, 2014), doi: 10.13140/RG.2.1.4564.8721; Gibbons, "Connected Infrastructure Activities."

16. California Institute for Energy and Environment and California Lighting Technology Center, UC Davis, "Campus-Wide Networked Adaptive LED Lighting," UC Davis website, 2014, https://cltc.ucdavis.edu/sites/default/files/files/publication/final_case-study-uc-davis-scaled-deployment-networked-ext-07-2014.pdf.

17. National Academies of Sciences, Engineering, and Medicine, *Solid-State Roadway Lighting Design Guide: Volume 2: Research Overview* (Washington, DC: Transportation Research Board, 2020), doi: 10.17226/25679.

18. Harsha Banavara, "Cybersecurity and Privacy for Connected Lighting," IES Street and Area Lighting Conference, Dallas, Texas, October 26, 2020.

19. Steve Taggert, "Chicago Smart Lighting Project," IES Street and Area Lighting Conference, Dallas, Texas, October 27, 2020.

20. Paul Tarricone, "It's 'Smart Cities,' as Sensor-Equipped Streetlights Offer the Proverbial Mother Lode of Data," Illuminating Engineering Society website, October 31, 2019, https://www.ies.org/lda/whats-the-word-on-the-street/?utm_source=IES&utm_medium=Email&utm_campaign=Client%20Updates&_zs=8K5NX&_zl=gjK32.

21. Los Angeles Bureau of Street Lighting, *LA Lights Strategic Plan 2020–2025* (Los Angeles: LABSL, 2020), http://bsl.lacity.org/strategic_plan.html.

22. Los Angeles Bureau of Street Lighting, *Strategic Plan*.

23. Examples include: Smartcitylab.com, smartcitiesdive.com, theagilityeffect.com, govtech.com, smartcityhub.com, smartcitiesworld.com.

24. Chris Danforth, "Utility-Led Smart City Attachments," IES Street and Area Lighting Conference, Dallas, Texas, October 27, 2020.

25. Maury Wright, "San Diego Broadly Deploys Cameras and Sensors on LED Street Light Poles," *LEDs Magazine*, February 20, 2019, https://www.ledsmagazine.com/leds-ssl-design/networks-controls/article/16695502/san-diego-broadly-deploys-cameras-and-sensors-on-led-street-light-poles-magazine.

26. Colin Santulli, City of San Diego Sustainability Department, Email, February 2, 2021.

27. Jesse Marx, "Smart Streetlights Aren't Delivering the Data Boosters Promised," Voice of San Diego website, April 29, 2020, https://www.voiceofsandiego.org/topics/government/smart-streetlights-arent-delivering-the-data-boosters-promised/.

28. Teri Figueroa, "Lawsuit Targets San Diego's Controversial Smart Streetlights," Government Technology website, December 17, 2019, https://www.govtech.com/smart-cities/Lawsuit-Targets-San-Diegos-Controversial-Smart-Streetlights.html.

29. Jesse Marx, "San Diego Can't Actually Turn Its Smart Lights Off," Voice of San Diego website, November 2, 2020. https://www.voiceofsandiego.org/topics/public-safety/san-diego-cant-actually-turn-its-smart-streetlights-off/; Teri Figueroa, "San Diego City Council Unanimously Backs Ordinances to Govern Surveillance Technologies," *San Diego Union-Tribune*, November 10, 2020, https://www.sandiegouniontribune.com/news/public-safety/story/2020-11-10/san-diego-city-council-unanimously-backs-ordinances-to-govern-surveillance-technologies.

30. "Seattle Information Technology: Surveillance Ordinance," Seattle Department of Information Technology, accessed March 15, 2021, http://www.seattle.gov/tech/initiatives/privacy/legal-protections_privacystatement.

31. Marx, "San Diego Can't Actually Turn Its Smart Streetlights Off."

32. Marx, "San Diego Can't Actually Turn Its Smart Streetlights Off."

33. Carly Minsky, "Internet of Things Help Cities Clean up Their Act," *Financial Times*, November 23, 2020, https://www-ft-com.ezp-prod1.hul.harvard.edu/content/4d8509e2-8c69-4cc5-b20d-3fb07a094467.

34. The Climate Group, "Connected LED Street Lighting Enable Smart Cities," https://www.theclimategroup.org/sites/default/files/downloads/tcg_smart_cities_introduction.pdf.

35. Ken Towsley, City of Syracuse Public Works Department, email, January 4, 2021.

36. "Massachusetts Town Integrates Road Condition Monitoring with Smart Street Light Deployment," *Smart Cities World* online, May 1, 2020, https://www.smartcitiesworld.net/news/news/massachusetts-town-integrates-road-condition-monitoring-with-smart-street-light-deployment-5245.

37. Richard Sarigumba, lead, LEDs/new lighting technology, Los Angeles Bureau of Street Lighting, Email, December 17, 2020.

38. Norma Isahakian, "Step by Step: An Update on the Quest for Smart-City Streetlights in Los Angeles," *Lighting Design & Application* 50, no. 7 (July 2020): 34–35.

39. Nate Berg, "Los Angeles Unveils the Sun-Blocking EV-Charging Streetlight of the Future," *Fast Company*, September 3, 2020, https://www.fastcompany.com/90546405/los-angeles-unveils-the-sun-blocking-ev-charging-streetlight-of-the-future.

40. Dave Pacetti, "Tampa Electric Gun Shot Detection," IES Street and Area Lighting Conference, Dallas, Texas, October 27, 2020.

41. Illuminating Engineering Society (IES), *RP 8-18: Recommended Practice for Lighting Roadway and Parking Facilities* (New York: IES, 2018), 2–9; Ronald Gibbons, "Connected Infrastructure Activities."

42. RP-8-18, 6-3.

43. John D. Bullough, "New Approaches to Lighting for Pedestrian Safety and Sense of Personal Security," Transportation Research Board Human Factors Workshop on *Walking at Night: The Pedestrian's Perspective*, January 8, 2017, https://www.pedbikeinfo.org/trbped/documents/2017/Bullough-TRBWorkshop-08Jan2017.pdf.

44. Ron Christian, Superintendent, City of Aspen Electric Dept., Email, December 18, 2020.

45. Naomi Fisher, "One Brush Stroke," *Lighting Design + Application* 50, no. 2 (February 2020): 20.

46. "Case Study: Cambridge Shared Streets," National Association of City Transportation Officials website, accessed April 20, 2021, https://nacto.org/case-study/cambridge-shared-streets/.

47. Ren21, Renewables in Cities: 2019 Global Status Report (Paris: UN Environmental Programme, 2019), https://www.ren21.net/wp-content/uploads/2019/05/REC-2019-GSR_Full_Report_web.pdf.

48. Ankit Gupta and Aditya Singh Bais, "Solar Street Light Market Size," Global Market Insights website, May 2018, https://www.gminsights.com/industry-analysis/solar-street-lighting-market.

49. Steve Dent, "Las Vegas Streetlights Are Powered by Your Footsteps," Engadget website, November 14, 2016, https://www.engadget.com/2016-11-14-las-vegas-kinetic-solar-streetlights.html.

50. "Solar Street Lights Chosen for Growing City," FirstLight Technologies website, March 14, 2019, https://www.firstlighttechnologies.com/solar-light-blog/solar-streetlights/.

51. "The Best LiFePo4 Battery for Solar Streetlighting," Greenpow Rechargeable Battery website, January 6, 2020, https://medium.com/battery-lab/the-best-lifepo4-battery-for-solar-street-light-c546f206c068.

52. Stephen Shickadance, "Here are Future Technologies for Solar Energy (And How They Fit into Lighting)," Greenshine website, accessed December 2, 2020, https://www.streetlights-solar.com/future-technologies-solar-energy-lighting.html.

53. National Renewable Energy Lab, *Lessons Learned: Community Solar for Municipal Utilities* (Golden, CO: NREL, December 2016), https://www.nrel.gov/docs/fy17osti/67442.pdf.

Chapter 7

Policies and Planning for Enhanced Lighting

7.1 Purpose and Scope of This Chapter

This chapter presents planning and policy tools to enhance lighting for pedestrians. There are numerous options for improving pedestrian safety and the pedestrian environment, but funding is scarce, so planners and policy makers need to consider how lighting enhancements can most effectively complement other potential improvements.

This chapter starts by describing warrants (criteria for determining whether street lighting is needed at particular locations). Then it covers citywide policy documents, recommending steps to develop a citywide or district lighting master plan or incorporating lighting recommendations in a Pedestrian Master Plan or similar document. Next, the chapter narrows to area and corridor plans, including how to integrate lighting with other improvements. While "lighting master plans" have often been conceived of as primarily technical documents for standardizing effective street lighting, the emphasis in this chapter is on broader planning that integrates lighting into other efforts to improve the pedestrian realm. Therefore, methods of inviting community input into the planning process are covered in detail. Funding and institutional issues also affect how lighting is treated as part of overall pedestrian improvement plans, and these are addressed in the final section.

With their greater flexibility, variability, remote control options, and potentials for "smart city" monitoring applications, LED lighting systems play a crucial role in the development of lighting master plans. The implementation of a successful plan is covered in Chapter 8, which focuses on integrating lighting into transportation design, operations, and maintenance.

7.2 Priorities and Warrants for New Street Lighting Installations

The highest priority for new or improved roadway lighting per IES *Recommended Practice RP 8*-18 are urban pedestrian crossing locations.[1] Individual locations often still need to be ranked by priority, using either warrants, comparison to light level standards, or by a customized scoring system. (The Seattle prioritization system is discussed in detail in the next section.) **Warrants** are criteria to

With funding scarce, lighting improvement locations can be prioritized by: warrants (standardized criteria for installation), comparison to lighting level standards, or by a customized scoring system. Scoring should take into account pedestrian volumes, traffic, and geometric features of the roadway along with characteristics of the surroundings.

determine when a new lighting installation is justified at a specific location, and they can be used to help prioritize locations. Meeting the warrants does not require installation. Factors outside of the warrant criteria may mitigate against the installation, such as cost or neighborhood opposition.

While the American Association of State Highway and Transportation Officials (AASHTO) has developed lighting warrants for freeways, highways, and bridges, neither AASHTO nor the Illuminating Engineering Society (IES) has developed their own warrants for arterial/collector/local streets, intersections, or walkways. (The IES recommended lighting levels can be used to determine where and how much lighting is needed for specific facilities.) The FHWA *Roadway Lighting Handbook* includes warrants for these facilities from the Transportation Association for Canada (TAC), but these are offered for possible consideration rather than as clearly recommended practice.[2]

Warrants can be useful in applications for federal funding for lighting improvements. A safety analysis and a study demonstrating the cost-effectiveness of the lighting project can also be a helpful complement to warrants. The AASHTO *Highway Safety Manual* may be used to frame such a study. This manual uses crash modification factors (CMFs) to help forecast the potential change in crash rate due to a proposed project or alternative treatments.[3]

The TAC warrant method for city streets is based on groups of geometric, operational, environmental, and crash factors. (See Table 7.1.) Each factor gets a 1–5 rating, and the ratings are weighted for their importance. The total point score indicates the need for lighting on the segment. The worst rating for each subsegment applies to the entire length of the road being considered.

For geometric factors, the weighted values are especially high for sharp horizontal curve radius, indicating it is an especially important criterion. For operational factors, the weighted value is especially high for pedestrian activity levels. For environmental factors, the weighted value is especially high for ambient lighting. For crash factors, the night/day crash ratio is the only factor listed on the form, but the total number of crashes should also be considered.

The TAC warrant method for intersections is based on geometric, operational, economic, and crash factors. Critical factors are traffic volumes and nighttime crashes. The total points indicate which is needed: full intersection lighting (uniform levels through the intersection), partial lighting (focused on conflict areas), or delineation (sufficient just to warn approaching traffic of the presence of an intersection). If the intersection is signalized, then full lighting should be provided. Otherwise, total points are used to determine the need. Cost/benefit analysis can be used along with the warrants for prioritizing locations.

TAC Warrants for Arterial, Collector, and Local Roads

Transportation Agency for Canada guidelines on when new roadway lighting is warranted. The US has no official national equivalent for streets. Courtesy of the Transportation Agency for Canada.

Road Name_____
From_____ to_____
City_____
Warrant Undertaken by_____
Company name_____
Date_____

Warrants for Lighting Arterial, Collector and Local Roads

Item No.	Classification Factor	Rating Factor 'R'					Weight 'W'	Enter 'R' Here	Score 'R' x 'W'
		1	2	3	4	5			
	Geometric Factors (See Note 6)								
1	Number of Lanes	≤ 4	5	6	7	≥ 8	0.15		
2	Lane Width (m)	>3.6	3.4 to 3.6	3.2 to 3.4	3.0 to 3.2	<3.0	0.35		
3	Median Openings/km	<2.5 or 1-Way	2.5 to 5.0	5.0 to 7.2	7.2 to 9.0	>9.0 or No Median	1.40		
4	Driveways and Entrances/km	<20	20 to 40	40 to 60	60 to 80	>80	1.40		
5	Horizontal Curve Radius (m)	>600	450 to 600	225 to 450	175 to 225	<175	5.90		
6	Vertical Grades (%)	<3	3 to 4	4 to 5	5 to 7	>7	0.35		
7	Sight Distance (m)	>210	150 to 210	90 to 150	60 to 90	<60	0.15		
8	Parking	Prohibited	Loading	Off Peak	One Side	Both Sides	0.10		
						Subtotal Geometric Factors			G
	Operational Factors								
9	Signalized Intersections (%)	80 to 100	70 to 80	60 to 70	50 to 60	0 to 50	0.15		
10	Left Turn Lane	All Major Intersections or 1-Way	Substantial Number of Major Intersections	Most Major Intersections	Half of Major Intersections	Infrequent Number or TWTL (See Notes 1 & 3)	0.70		
11	Median Width (m)	>10	6 to 10	3 to 6	1.2 to 3	0 to 1.2	0.35		
12	Operating or Posted Speed (km/h) (See Note 5)	≤ 40	50	60	70	≥ 80	0.60		
13	Pedestrian Activity Level (See Note 2)			Low	Medium	High	3.15		
						Subtotal Operational Factors			O
	Environmental Factors								
14	Percentage of Development Adjacent to Road (%) (See Note 4)	nil	nil to 30	30 to 60	60 to 90	>90	0.15		
15	Area Classification	Rural	Industrial	Residential	Commercial	Downtown	0.15		
16	Distance from Development to Roadway (m) (See Note 4)	>60	45 to 60	30 to 45	15 to 30	<15	0.15		
17	Ambient (off Roadway) Lighting	Nil	Sparse	Moderate	Distracting	Intense	1.38		
18	Raised Curb Median	None	Continuous	At All Intersections (100%)	At Most Intersections (51% to 99%)	At Few Intersections (≤ 50%) (See Note 7)	0.35		
						Subtotal Environmental Factors			E
	Collision Factors								
19	Night-to-Day Collision Ratio	<1.0	1.0 to 1.2	1.2 to 1.5	1.5 to 2.0	>2.0 (See Note1)	5.55		
						Subtotal Collision Factors			A

G + O + E + A = Total Warranting Points

Warranting Condition	60.00
Difference ±	-60.00	D

Notes:
1 Lighting Warranted
2 Pedestrian Activity Level (Refer to 9.1.3 – Pedestrian Related Definitions)
3 Two-Way Left Turn Lane
4 Development Defined as Commercial, Industrial or Residential Buildings
5 85th Percentile Night Speed Should Be Used if Available, Otherwise Posted Speed Shall Be Used
6 Worst Case Geometric Factors for a Segment of Roadway Shall Apply
7 Also Includes Isolated Medians (Non-Continuous) Between Intersections.

v1.0

7.3 Citywide Policies and Policy Documents

Citywide policies for lighting are occasionally provided in **citywide lighting master plans**. Seattle and San Jose plans are summarized here because of their focus on pedestrian needs. Other US cities have also developed broader citywide lighting master plans, including: Evanston, Illinois; Galveston, Texas; Salt Lake City, Utah; Columbus, Ohio; and Burbank, California. Cities outside the U.S. such as Lyon, France, Singapore, and Barcelona, Spain, also have produced notable lighting plans. (Lyon's is discussed in detail in Chapter 9 regarding its attention to historical lighting.) Lighting policies also are provided in broader pedestrian **policy documents**, such as San Francisco's *Better Streets Plan*, and in more focused lighting-specific policy documents, such as the San Francisco Public Utilities Commission catalog of approved pedestrian-scale lighting equipment.

Strategic plans, such as the Los Angeles Bureau of Street Lighting *Strategic Plan*, address the mission and key initiatives of an agency. While such strategic plans often cover policies of special interest to the pedestrian environment, because they focus so broadly on organization and partnerships, they are not covered in detail here.

7.3.1 Citywide Pedestrian Lighting Plans

A citywide or district pedestrian lighting plan is a valuable tool to enhance safety, security, and livability. Such a plan can be a separate document, or (especially for small and medium-size cities) it may be a portion of a broader pedestrian master plan or overall lighting plan. The following recommendations are based especially on the Seattle and San Jose examples highlighted below.

Plan contents focus on priority locations for new or enhanced lighting, recommended light levels, measures to reduce adverse impacts, institutional responsibilities, and funding. The plan should be future-oriented, formulated to respond to trends and accommodate growth and development. (Recommended contents of the plan are listed in Table 7.2.)

The Seattle Department of Transportation developed a citywide plan focused on pedestrian-scale lighting in 2012, covering policy and regulations, planning and design, and funding.[4] The lighting plan was intended to help implement Seattle's *Pedestrian Master Plan*. It provides a blueprint for the needs and opportunities for pedestrian lighting for public walkways and gathering areas. The citywide pedestrian lighting plan fleshes out several important components of the broader *Pedestrian Master Plan*: addressing goals and objectives

The development of an effective citywide lighting plan requires the involvement of a broad range of government agencies, residents, and community groups. The plan needs to be useful several years after its publication, thus requiring consideration of such factors as technology improvements, ongoing changes in transportation policies, as well as environmental and economic goals.

Recommended Contents for a Citywide Lighting Plan

1. Purpose
2. Development of the plan, including community participation
3. Data about existing community characteristics and infrastructure
4. Existing lighting inventory
   o Locations
   o Sources
   o Light levels
   o Design criteria
5. Goals
6. Problem analysis and recommendations related to:
   o Safety
   o Security
   o Aesthetics
   o Lighting design criteria
   o Potential adverse impacts (e.g., energy use, light pollution, including skyglow affecting observatories)
   o Economic development
   o Smart city applications
   o Maintenance
7. Implementation plan, including:
   o Costs and funding
   o Responsibilities
   o Future evaluation

related to safety and walkability, helping prioritize pedestrian improvements based in part on after-dark conditions, supporting the development of cost estimates, and ensuring that street designs incorporate pedestrian-scale lighting into the limited street right-of-way.

Methods for prioritizing lighting locations should rank virtually all locations allowing pedestrians. Example methods include Seattle's pedestrian lighting needs scores and San Francisco's WalkFirst system for rating locations for a range of potential improvements. Potential factors for lighting need scores include the following (assuming the typical case where comprehensive data on pedestrian volumes are not available):

- Transit proximity (as a factor in pedestrian volumes)
- Street classification (as a factor in vehicle and pedestrian volumes)
- Land use classification (as a factor in pedestrian volumes)
- Crime levels in low light conditions
- Collision record, particularly low light conditions
- Complex intersections or other visually challenging locations
- Potential for coordination with other projects
- Citizen input (including complaints about specific locations and response to surveys)

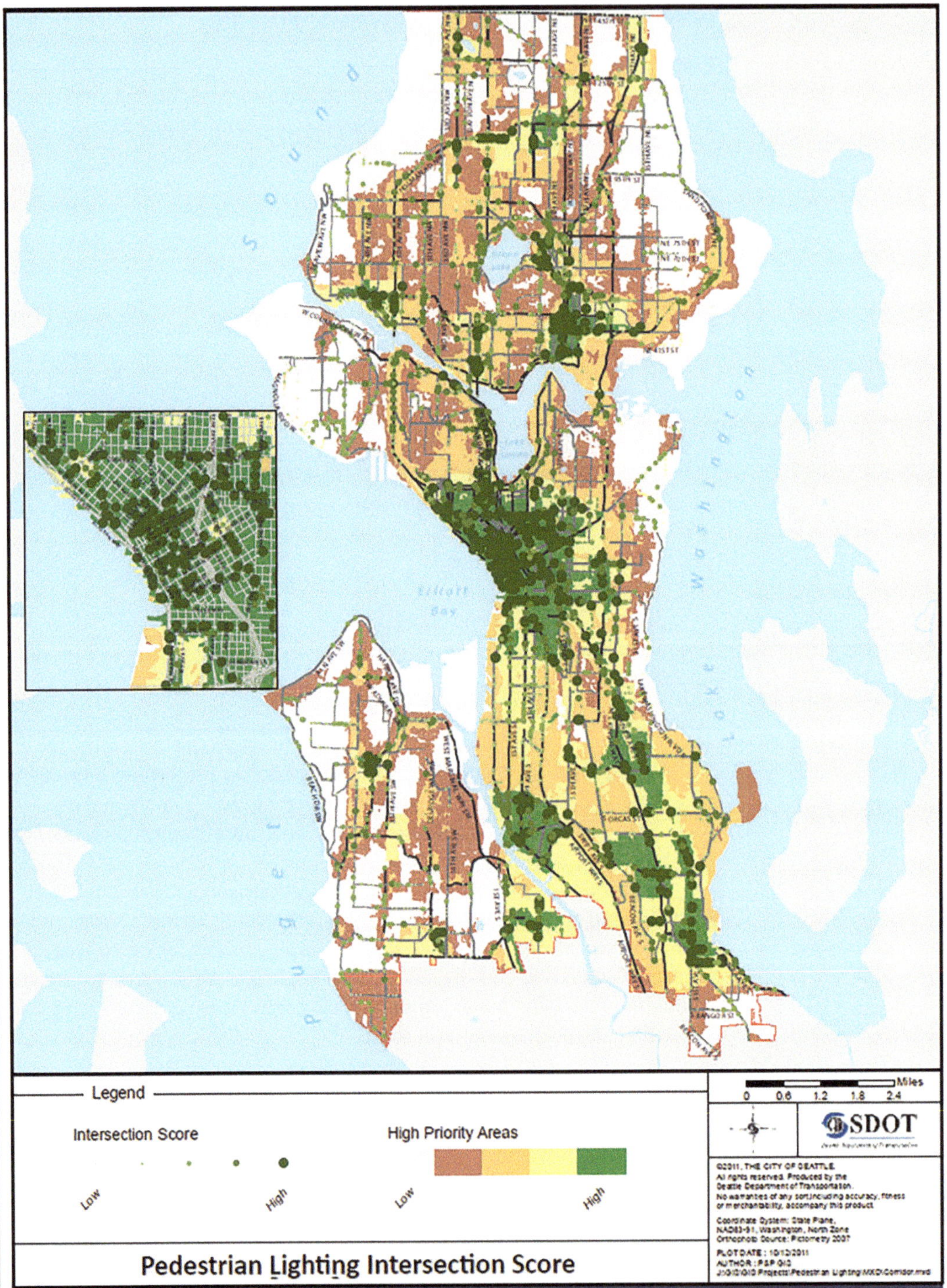

22

June 2012

Legend

Intersection Score

Low   High

High Priority Areas

Low   High

Miles
0   0.6   1.2   1.8   2.4

SDOT

©2011, THE CITY OF SEATTLE.
All rights reserved. Produced by the
Seattle Department of Transportation.
No warranties of any sort, including accuracy, fitness
or merchantability, accompany this product.

Coordinate System: State Plane,
NAD83-91, Washington, North Zone
Orthophoto Source: Pictometry 2007

PLOT DATE : 10/12/2011
AUTHOR : P.SP GIS
J:\GIS\GIS Projects\Pedestrian Lighting\MXD\Corridor.mxd

Pedestrian Lighting Intersection Score

Map 4: Pedestrian Lighting Intersection Score

This scoring system prioritizes locations expected to have the highest number of pedestrians and where nighttime pedestrian safety and crime levels are a concern. However, minimum illumination levels should be provided at all locations with need, based on the IES standards. Locations that do not need to be considered for supplemental pedestrian lighting include isolated industrial or warehouse districts and other areas expected to get negligible pedestrian use (e.g., the sides of streets abutting freeways or steep, undeveloped hillsides).

Seattle developed priorities for areas, arterial segments, intersections, and pathways. (See Figure 7.1.) Area prioritization criteria in the Seattle Plan include:

- Pedestrian demand (based on land uses)
- Socio-economic status (with historically under-served areas prioritized)
- Street segment classification (importance in the pedestrian network)
- Crime hotspots

Arterial corridors were scored based primarily on the area prioritization, taking into account also the cost-effectiveness of potential construction projects. Then intersection scores were developed, adding in pedestrian safety data. Finally, stairs and pathways were also prioritized.

San Francisco's innovative WalkFirst project ranked roadway lighting improvements as high effectiveness, medium cost, and long time frame.[5] The project's toolkit webpage suggested that lighting improvements should be targeted especially to locations with a high nighttime crash profile and to complex intersections.

Light source technology, pole/luminaire style, and smart lighting should be addressed, reflecting technology changes. Typically, this will involve consideration of how to take advantage of LEDs and smart lighting potentials for greater control and networking of illumination (as discussed in detail in Chapter 6.) Issues beyond illumination must be analyzed when considering sensors and services that can be hosted on smart poles.

San Jose, California, revised its *Public Streetlight Design Guide* in 2016 to respond to City Council policy allowing broad spectrum (white) lighting (LED, plasma, and induction) due to its greater energy efficiency and longer lifetime than existing lighting (typically high and low-pressure sodium).[6] City Council policy also allows dimming street lights during times of lower pedestrian and vehicle volumes.

San Jose's *Downtown Street and Pedestrian Lighting Master Plan* was amended in 2017 to respond to technology changes including LED and adaptive lighting.[7] The Plan provides maps of recommended street and pedestrian light source types, including separate maps for historic and contemporary pedestrian-scale lighting. (See Figure 7.2.)

Limiting light pollution is another key focus. The San Jose *Downtown Street and Pedestrian Lighting Master Plan* aimed to minimize skyglow impacts on the Lick Observatory (several miles from the urbanized core of the city). Recommendations to accomplish this include shielding to block uplight, BUG limits, timing guidelines to reduce light levels at lower-need times, and consideration of dynamically adaptive lighting.

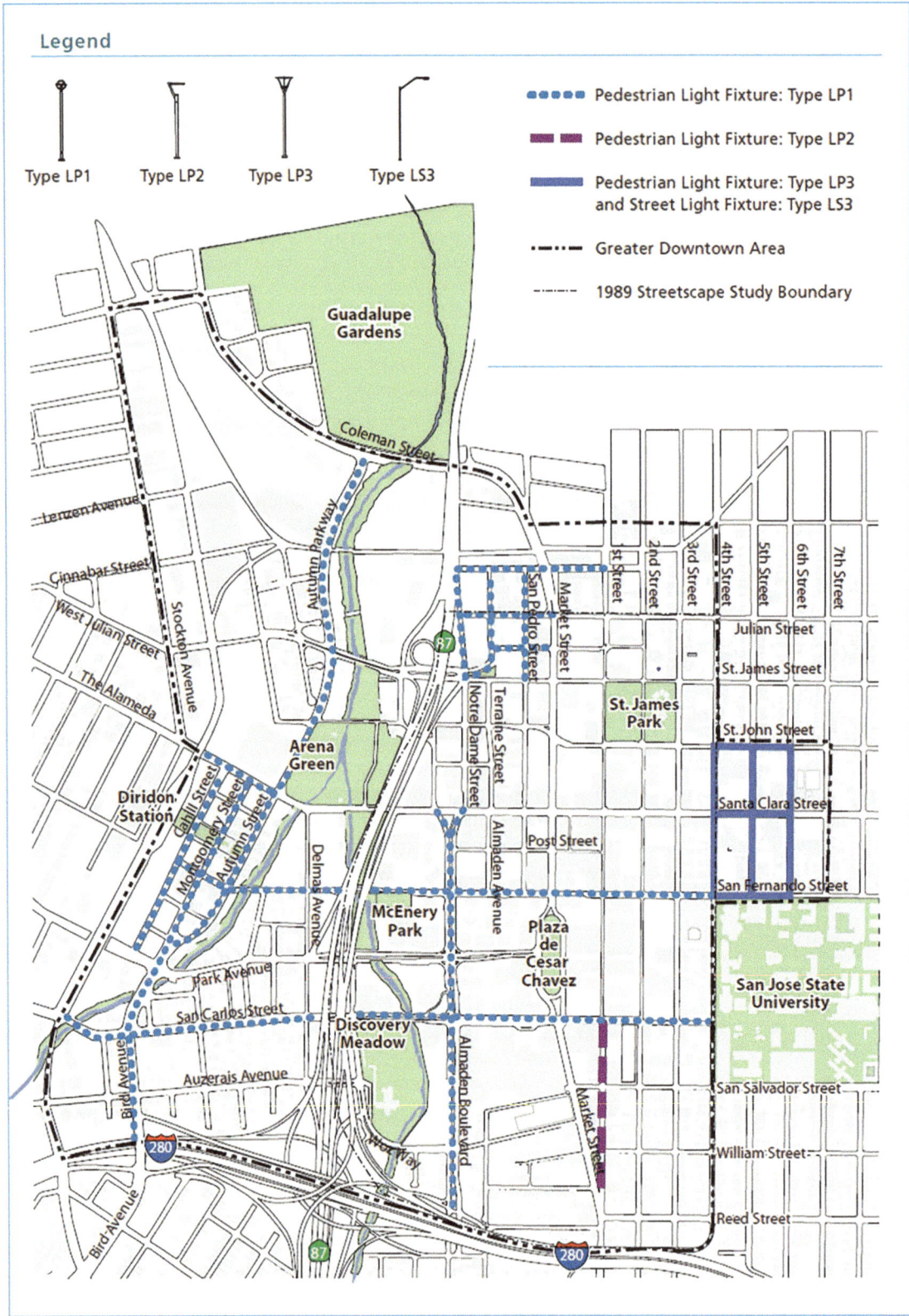

Legend

Type LP1   Type LP2   Type LP3   Type LS3

●●●●● Pedestrian Light Fixture: Type LP1

▬ ▬ ▬ Pedestrian Light Fixture: Type LP2

▬▬▬ Pedestrian Light Fixture: Type LP3 and Street Light Fixture: Type LS3

—··—··— Greater Downtown Area

— — — 1989 Streetscape Study Boundary

Map 6: Recommended Contemporary Street and Pedestrian Light Fixtures

◀ Figure 7.2

Downtown San Jose Pedestrian-Scale Lighting Recommended Locations. Map shows recommended locations for "contemporary" pedestrian and street lights. There is a separate map for "historic" lighting. Courtesy of City of San Jose.

Funding and institutional responsibilities are key topics for lighting plans. For example, Seattle's *Pedestrian Lighting Plan* states:

> Recognizing that pedestrian lighting is a key element for the safety and security of pedestrians and is a principal tool to encourage walking as transportation, the operations and maintenance of pedestrian lights should be fully funded through the City's General Fund.[8]

Additional detail on funding is provided below.

The recommended ***steps to develop the plan*** include:

1. Inventory existing lighting fixtures
2. Analyze after-dark crash patterns and locations
3. Assess citizen preferences for lighting fixtures and strategies, including field surveys after dark and discussions of "smart lighting" options
4. Inventory other crash, road, and environmental data (to help understand lighting needs relative to other pedestrian needs)
5. Adopt a scoring system for determining lighting need
6. Map lighting needs by priority classification
7. Prepare benefit/cost analysis for broad alternative lighting and smart city application strategies
8. Conduct a benefit/cost analysis for prioritizing lighting improvements along with other potential roadway improvements and compare also on metrics such as:

   - ease of implementation,
   - time frame,
   - funding availability

9. Recommend a specific program of improvements
10. Develop a funding and implementation strategy
11. Prepare for adoption by a policy body through public review of a draft plan and potentially environmental review

A pedestrian lighting plan's development can be led by the city's transportation department or unit. This allows it to be integrated with other transportation planning efforts. Agencies or utilities that maintain and operate street lighting may also lead plan development and should certainly play a key role. Other key departments and organizations should be actively involved, including:

- Public Works (if a separate department)
- City Planning
- Transportation funding agency or authority (if appropriate)
- Information technology (to address smart lighting options)
- Citizen advocacy and non-profit organizations
- Business representatives

7.3.2 Pedestrian Realm Master Plans

Many communities have developed citywide plans for the pedestrian realm, focusing on a combination of transportation and urban design needs. Lighting should be an essential component of these documents. However, these plans cannot typically treat lighting in sufficient depth, so a citywide lighting master plan is usually still desirable.

A Pedestrian Master Plan (PMP) should establish the importance of pedestrian-scale lighting and illumination of crossings. Seattle's 2017 PMP does this, noting that lighting was one of the more frequently mentioned improvements requested at community meetings.[9] The PMP recommends updating the *Pedestrian Lighting Citywide Plan* to address new funding needs and lighting technologies. It also recommends that the City's *Streets Illustrated: Right-of-Way Improvements Manual* require pedestrian-scale lighting downtown and that it specify light levels and pole spacing. The 2019 edition of that Manual addresses light levels and pole spacing.[10] It explicitly supports the prioritization of locations for pedestrian lighting from the 2012 *Pedestrian Lighting Citywide Plan*.

A PMP or pedestrian realm plan needs to treat lighting in conjunction with other aspects of urban design. For example, the award-winning San Francisco *Better Streets Plan* (BSP) provides general priorities for pedestrian-scale lighting according to street type. The BSP prioritizes the following locations for pedestrian-scale lighting:

- Streets with high pedestrian volumes
- Key civic, downtown, and commercial streets
- Streets with concerns about pedestrian safety and security, such as at freeway underpasses
- Small streets such as alleys and pedestrian pathways

Coordination with other elements of urban design, such as street trees, is a key feature of such a plan. For example, the BSP offers general design guidelines (pole spacing, location, coordination with tree locations), discussing the desirable light distribution through the full street right-of-way.

These plans should point to established standards for light levels. For example, the BSP supports IES recommendations for light levels for roadways, intersections/crosswalks, and sidewalks. It also addresses correlated color temperature (CCT).

Controlling costs and adverse impacts, funding, and operational procedures are other important considerations. The BSP addresses energy efficiency, equipment procurement, and maintenance. The BSP also mentions lighting districts as one method of funding for the installation of pedestrian-scale lighting.

7.3.3 Citywide Lighting Equipment Standards

Some cities publish street light equipment standards or catalogs to promote quality lighting, minimize financial costs and adverse impacts, and to support efficient installation and maintenance. These should specifically address pedestrian-scale lighting and historic/ornamental lighting. They may be included in broader policy documents or in a stand-alone document.

In 2018 the San Francisco Public Utilities Commission (SFPUC) issued a policy statement supporting increased use of pedestrian-scale lighting, but only with approved equipment listed in a catalog.[11] The City and County of San Francisco will pay for the increased electricity costs and maintenance of pedestrian-scale lighting installed by agencies following these policies.

The catalog initially listed ten pedestrian luminaires that meet the illumination and uniformity standards of IES RP-8-18. The SFPUC also issued guidance on using luminaires and poles not in the catalogue. This is permissible if demonstrated that the variance is not materially detrimental to public welfare, public finances, SFPUC maintenance and operations, or energy efficiency. An overstock supply is required.

Recommended street lighting equipment may also be included in broader lighting plans. For example, the San Jose *Downtown Street and Pedestrian Lighting Master Plan* lists and illustrates poles and luminaires for specific street types, including acceptable models and their BUG ratings.

7.4 Area and Corridor Planning Projects

Pedestrian-oriented plans for areas and corridors may focus on transportation, economic development, aesthetics, or a comprehensive consideration of these topics. Lighting enhancements should be considered as part of these planning projects, ideally drawing on citywide lighting plans described above. These area or corridor plans may also provide excellent opportunities for innovative pilot projects that can be evaluated for potential use in other locations.

7.4.1 Integrating Lighting Improvements into Area/Corridor Plans

Lighting for an area or corridor should be considered along with the numerous other factors that affect the pedestrian environment. Examples discussed below from San Francisco and Seattle illustrate important features of area/corridor plans:

- Community involvement
- Inter-agency coordination
- Needs analysis
- Integration of lighting and other improvements
- Consideration of new technology.

Lighting enhancements should be considered for any major pedestrian or complete streets project or plan for a large area or corridor. For example, the City of Seattle includes a lighting analysis for any corridor improvement project with a construction cost of over $100,000.

There are excellent guides to prioritizing a broad range of investments in improving the pedestrian environment. One manual on *How to Develop a Pedestrian Safety Action Plan* recommends a comprehensive strategy for prioritizing improvements for specific locations.[12] San Francisco's WalkFirst project combined public input with statistical analysis of pedestrian injury patterns and cost-effectiveness data on engineering measures to develop a recommended program of improvements for specific locations.[13]

The classic steps for an area or corridor transportation plan include the following:

1. **Problem identification** is based on focused data collection and community input. Data collection should include a review of lighting conditions for pedestrian injuries and fatalities, plus an inventory of lighting in the planning area. Ideally, light levels should be measured and compared to standards.
2. **Plan goals** should include reducing pedestrian injuries, including nighttime injuries specifically, and improving the pedestrian environment. "SMART" objectives for each broad goal should provide Specific, Measurable, Achievable, Relevant, and Time-Based targets where feasible.[14]
3. **Alternatives development** describes and quantifies packages of improvements (and supporting policies) to achieve goals and objectives. For lighting, alternatives could include enhancements to conventional street lights, pedestrian-scale lighting, and more technologically advanced options such as adaptive lighting and smart city applications. The federally developed PEDSAFE system includes free online software that allows the user to match countermeasures, including lighting, to 12 crash types or eight performance objectives.[15]
4. **Alternatives evaluation** is based on criteria such as expected benefits, costs, other adverse impacts, implementation time, and difficulties.
5. **Draft plan recommendations** are formulated primarily for plan review and approval. At this stage, sufficiently detailed information should be developed to allow for environmental review and policy board approval.
6. **Plan approval** is provided by the policy board(s), such as a City Council, Planning Commission, Transportation Commission, or other bodies. These bodies may approve with specific conditions, e.g., reducing the geographic scope of the project or requiring environmental mitigation measures.
7. **Implementation** of lighting recommendations may be carried out by a different agency or utility from the agency managing the plan or project. The implementing agency may develop a detailed design after plan approval, but the planning or managing agency needs to be involved to determine consistency with approvals and environmental commitments.
8. **Evaluation of the effectiveness of the project** is desirable to refine the plan for the specific area/corridor, to ensure proper maintenance or corrective measures are undertaken, and as input for later efforts.
9. **Community involvement** is addressed in more detail below.

Multiple complementary planning efforts and different agencies may need to address lighting in a single area. For example, the Balboa Park area of San Francisco hosted two recent projects to enhance lighting, primarily to improve access to public transit. The Balboa Park neighborhood includes the most heavily used public transit transfer hub in the San Francisco Bay Area outside of a major downtown district. It includes a BART subway station (serving that hybrid commuter rail and urban rapid rail system), San Francisco Muni surface light rail lines, extensive Muni bus service, and commuter/community shuttle stops.

The San Francisco Bay Area Rapid Transit District constructed a walkway between its station and the community college and park, as well as improving plazas on its properties. This project included landscaping and lighting. In 2012 the San Francisco Municipal Transportation Agency (SFMTA) also developed and implemented a station area capacity enhancement plan, which included a range of pedestrian safety and amenity measures.[16] The transportation agencies coordinated with each other and with numerous other stakeholders.

When transit hub users were questioned about station area transportation needs, lighting ranked as the third most significant barrier to transferring.[17] The consulting firm Jacobs Engineering observed light levels qualitatively to identify specific problem areas where pedestrian-scale lighting would be helpful. Pedestrian-scale lighting was recommended for every 50 feet to supplement street lights that were typically 100 feet apart. The *Station Capacity Plan* also referred to the potential for artistic lighting to improve neighborhood identity, but it did not provide specific recommendations.

7.4.2 Translating Plan Recommendations into Design

It is crucial that lighting recommendations be structured by an area or corridor plan to advance into design. The plan can facilitate this by identifying and supporting funding and connecting planning, design, and implementation agencies.

For example, after completing the *Station Capacity Study*, which recommended pedestrian-scale lighting at specific priority locations in the Balboa Park Station Area, the SFMTA obtained funding and worked with lighting engineers for San Francisco Public Works and the Public Utilities Commission to design and construct the pedestrian-scale lighting improvements. Transitional, limited street lighting installations were made to fill major gaps before the full pedestrian lighting could be added.

The pedestrian-scale lights selected were LED lights on 12-foot poles. (See Figure 7.3.) The LED lighting is highly efficient, at 117 watts. The correlated color temperature is about 4,000 degrees Kelvin, providing a relatively bright white light with good color rendition, consistent with the VTTI recommendations described earlier.

The San Francisco Balboa Park project suggests how standard street lighting and pedestrian-scale lighting improvements can be integrated. It also exemplified the inclusion of lighting improvements into a broader area transportation plan.

San Francisco Western Addition Community Transportation Plan

San Francisco's Western Addition is an historically underserved Community of Concern (CoC) with a high concentration of low-income housing and a large proportion of minority residents. As a legacy of urban renewal projects in the 20th century, the neighborhood has wide streets and one-way streets that encourage high vehicle speeds.

The SFMTA's *Community-Based Transportation Plan* aimed to improve transportation, with an emphasis on walking, bicycling, and public transit. Near-term, medium-term, and long-term recommendations were developed. The plan was prepared with extensive community involvement, using a community partner, Mo'MAGIC.

The need for additional street lighting was one of the primary concerns raised by community members. Some 80 percent of survey respondents felt more street lighting was needed for pedestrians. Lighting was one of the top four requested improvements.

Evaluation of potential improvements combined technical analysis and community input. Although a transportation-focused study, the project addressed other community issues like crime and sense of place. In working with the community on possible solutions, planners used the Crime Prevention through Environmental Design (CPTED) approach, which includes landscaping, fencing, street lighting, and other

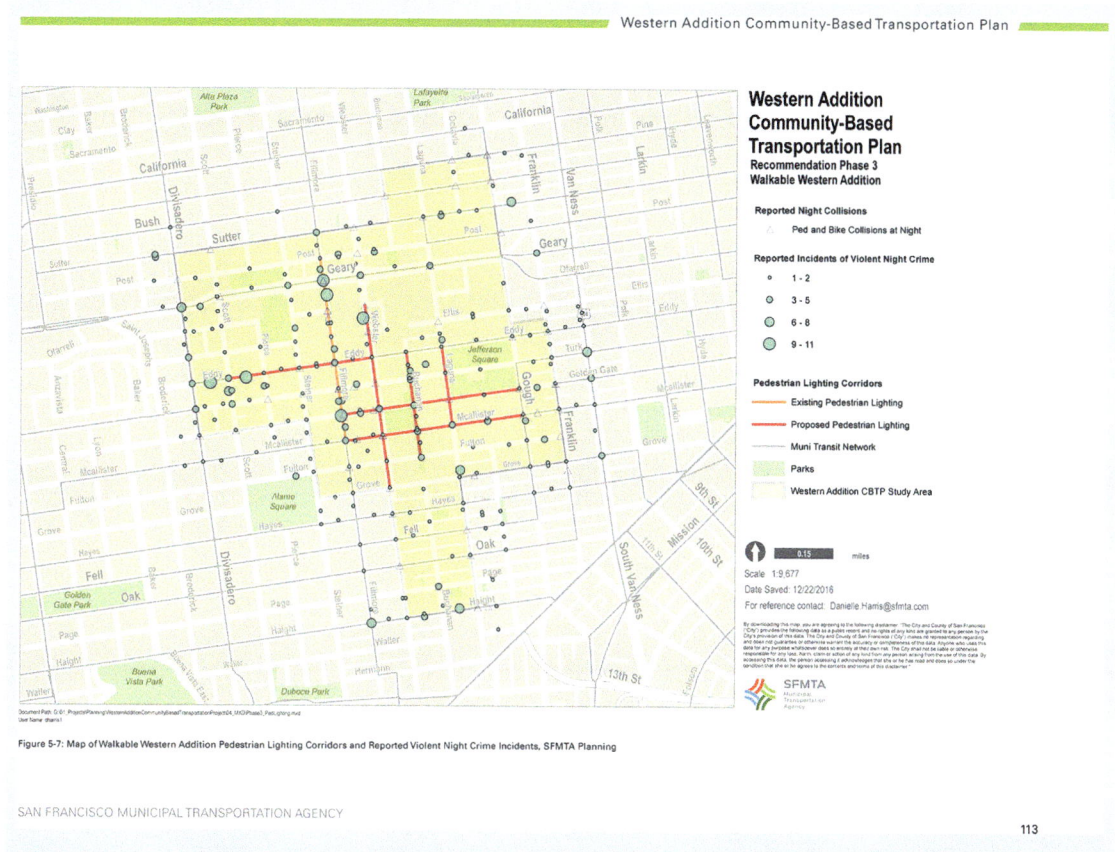

Figure 5-7: Map of Walkable Western Addition Pedestrian Lighting Corridors and Reported Violent Night Crime Incidents, SFMTA Planning

SAN FRANCISCO MUNICIPAL TRANSPORTATION AGENCY

113

▲ Figure 7.4

Western Addition (San Francisco) Pedestrian-Scale Lighting Recommended Locations. Courtesy of San Francisco Municipal Transportation Agency. Map produced by Danielle Harris.

features.[18] The resulting recommendations included pedestrian-scale lighting on several corridors, based on nighttime pedestrian crash records, nighttime violent crimes, and Muni bus routes. (See Figure 7.4.)

7.5 Community Involvement and Equity

Community involvement starts with local government attention to resident and business concerns before there is a specific plan or project. It then includes structured use of stakeholder information, ideas, and preferences during the development of formal plans. And it continues with a consultation during and after plan implementation.

Local governments need a proactive approach to assessing and responding to citizen complaints. For example, 311 call records can be analyzed to determine where citizens have identified locations of repeated concerns.

Equity in transportation and roadway lighting is an increasing concern that should help shape community involvement programs and transportation/lighting plans and policies. The *Portland Plan*, for example, provides a framework for equity as a core value built into all its strategies.[19]

7.5.1 Community Preferences for Lighting in Area/Corridor Plans

Obtaining effective community input in planning studies is challenging. Participation must be convenient for citizens. Lighting needs may be overlooked because they're most apparent after dark when agency staff or some stakeholders are less likely to be present in the study area. Therefore, it is important to consider evening walk audits or other tools to ensure that plan recommendations cover low-light needs.

California's Local Government Commission prepared a public involvement guide that covers methods for attracting robust participation, types of meetings and processes, and meeting/field exercises.[20] Besides walking audits, it describes several other participatory exercises such as focus groups and photo visioning (photo comparisons). Consultants can provide customized photo transformations of streetscapes. Example streetscape images online allow the addition of lighting concepts.

Another tool for resident input, pedestrian facility quality scales, have been developed to assess perceptions of walking safety and comfort. The San Francisco Department of Public Health devised a Pedestrian Environmental Quality Index, or PEQI, which has been used in studies in San Francisco and other cities.[21] The PEQI considers five elements: (1) intersection safety, (2) traffic features, (3) street design features, (4) land use, and (5) perceived safety. Observers are to note the presence of pedestrian-scale lights at an intersection, without requiring further detail on the effectiveness of lighting. The PEQI does not ask for a qualitative or quantitative assessment of light levels, the physical attractiveness of light fixtures, or similar factors.

San Francisco's *Western Addition Community-Based Transportation Plan's* extensive community outreach program provided examples of how to address lighting in an areawide complete streets plan.[22] For example, a questionnaire

As examples of innovative techniques to gather community input: residents can rate "future" streetscape images with alternative lighting schemes, questionnaires can ask for specific locations for enhanced lighting, and walk audits can be conducted during after-dark conditions (and looking at lighting options). Also, lighting problems for pedestrians can be reported online even by those who find meeting attendance difficult.

used at meetings and online asked about the "need for more street lights." Planners used a Design Game and Toolkit, asking meeting attendees and others to identify specific locations for enhanced lighting. (Boston's Beta Blocks program has used a similar game approach at meetings to obtain community preferences for smart lighting and other smart city technology, promoting family engagement.[23] The Beta Blocks program also sponsored a "robot block party" that attracted 4,500 participants to learn hands-on about artificial intelligence, automated vehicles, and robotics – an approach which could certainly be extended to include smart lighting.)

The Western Addition community outreach program was developed with Mo'MAGIC, a collaborative San Francisco neighborhood-based nonprofit organization, which has developed partnerships with community organizations around social justice. The office of then-County Supervisor London Breed also provided input into the study.

7.5.2 Community Preferences in Focused Lighting Plans

Planning projects focused exclusively on lighting can use the public participation tools mentioned above. They also can benefit from focused measures that may be difficult to include in broader studies, such as field demonstrations of different lighting technologies and questionnaires on lighting preferences.

In its co-authored guidebook for New York State, the International Nighttime Design Initiative (NTD) encourages municipalities to obtain community input in their smart lighting programs, building citizen knowledge of new lighting capabilities.[24] The book posits that engagement can range from awareness-raising to participation in decisions for area applications such as light level setting, color tuning, and time-based on-off dimming and switching. An engagement activity was conducted during the book's research phase. In Saratoga Springs, New York, a NightSeeing™ event was hosted to both educate and elicit stakeholder input. The activity included an after-dark talk and a walk around downtown Saratoga Springs to discuss public and private lighting, providing immediate feedback to the study authors and a project advisory committee. Subjects included shop windows, safety, ambiance, and light pollution.

For an IES conference poster, NTD described novel, in-depth community interaction. This included developing a "responsive lighting testbed" and a "responsively-lit outdoor patch" with the aim to gather public and technical evaluation data on lighting concepts.[25]

The San Jose *Citywide Street Light Plan* and the *Downtown Pedestrian Lighting Plans* used a demonstration and community field test conducted by Clanton & Associates and Dr. Ronald Gibbons. The study systematically compared the performance of different streetlight technologies – LPS, HPS, induction, and LED – at full brightness and dimmed approximately by half. Community members were surveyed about their response to the lights. A small target visibility study determined how well people could detect objects at different lighting levels. The study helped establish the parameters for when, where, and how much lights would be dimmed.

The University of Utah hosted an annual "Walk After Dark."[26] Organized by the University's Occupational & Environmental Health and Safety Office, partnering with the Associated Students organization and Facility Operations, the survey, held on an evening of optimal darkness, identified broken lights, dark areas, damaged sidewalks, and other barriers to nighttime travel.

In 2018 nearly 140 volunteers used a phone app to geolocate problems. This effort has also been useful for the campus lighting plan, which aims to enhance light levels on campus while minimizing light pollution. (The University website features information on Walk after Dark and campus lighting. It also provides a link specifically to report lighting problems at any time.[27])

Researchers at the Swedish Lund University used a mocked-up pedestrian path to collect data on pedestrian response to three different light sources.[28] Researchers observed: walking speeds; detection of obstacles, facial expressions, and street signs; emotional response; and subjective evaluation of lighting quality. They also made objective comparisons of lighting features. Although an LED application (with higher power and illuminance, and CCT of 3,800 degrees Kelvin) was found best for measured detection distances and perceived visibility, it also was considered "less pleasant" than other light sources.

An example of a focused community involvement effort in response to neighborhood complaints was provided by the university community of Davis, California.[29] In 2014, the city converted sodium-vapor street lights to LED lights. The new lights had a cooler, whiter color (CCT of 4,000 Kelvin) compared to the warmer, amber (2,200 Kelvin) old lights. Resident complaints led the city council to stop installation after 1,400 lights were retrofitted. The complaints alleged the new fixtures were too bright, with excessive glare and light pollution.

The city then sponsored a field demonstration of several options. The city settled on a 2,700 Kelvin LED model, replacing 650 of the retrofitted versions in residential areas and adding new locations. Shielding was also improved.

7.5.3 Equity in Lighting

Lack of equity in transportation facilities is a contributor to higher pedestrian injuries and fatalities in lower income neighborhoods and communities of color, which are often more dependent on walking for basic transportation. For example, Native Americans are nearly three times more likely than non-Hispanic whites (and four times more likely than Asian/Pacific Islanders) to be killed while walking.[30] Those living in areas with median household income of $3,000–$36,000 have a pedestrian fatality rate three times higher than areas with median household income of $79,000–$250,000.

Some cities have explicitly addressed equity considerations in lighting projects and programs. For example, the City of Seattle developed an Equity Impact Analysis tool and used it to assess the process for relamping street lights. Seattle City Light relied on reports of outages to prioritize relamping, but it found that these were less likely to be generated by low-income and minority neighborhoods, especially with recent immigrants. So Seattle started using the age of luminaires to prioritize, focusing especially on South Seattle, which had higher low-income and minority representation. More recently, Seattle City Light moved to use a smart photocell that could detect and flag outages equitably.

The City of Portland, Oregon, planned an LED conversion project to use equity considerations to prioritize retrofits. The City's Bureau of Transportation in 2013 partnered with the non-profit Coalition for a Livable Future to develop a priority scoring system that considered demographics, safety and security data, and transportation network access. As discussed earlier, the San Francisco Western Addition Community-Based Transportation Plan was developed with a community organization, Mo' MAGIC, with deep ties with minority neighborhood groups.

People with disabilities are also a critical stakeholder group for community involvement. Both visual impairments and mobility challenges benefit from special lighting strategies. Although someone who is completely blind does not personally need lighting to see, it is important for everyone to be visible to others, and any seeing eye dogs or sighted companions need light. Those with reduced vision often require higher levels of illumination. Someone using a walker or wheelchair may also be more susceptible to tripping hazards or obstructions. In developing its lighting plan, Lyon (France), sought to consult people with disabilities.

7.6 Funding

It is essential that transportation and lighting plans provide customized funding strategies. Both capital costs and operations/maintenance costs need to be considered. A funding element identifies the cost stream, the potential funding sources, and issues such as eligibility or competitiveness for each funding source.

7.6.1 Funding Considerations

Capital improvement projects, such as the installation of new or improved lighting, often require a complicated, multi-source funding stream. Different funding sources may be used for planning, design, and installation phases. During the planning and design phases, it is important to assess whether it will be feasible to obtain funding for later phases and whether eventual operations and maintenance can also be funded.

Lighting plans and projects need to address funding requirements, including not just capital costs, but also operations and maintenance needs. Cities can consider innovative sources, such as developer contributions and public-private partnerships.

New lighting installations may trigger increased needs for:

- Electricity
- Electrical maintenance, including preventive maintenance and problem response
- Tree pruning
- Public outreach (to address community concerns)

Lighting projects (or lighting enhancements as part of broader transportation projects) should not be assumed to be eligible for funding from general transportation sources. For example, San Francisco's transportation sales tax (Prop K) prohibited its use for lighting projects. The Illinois Transportation Enhancement Program (using federal funding) considered stand-alone street or pedestrian lighting projects ineligible, except for street lighting in historic districts. Such prohibitions may be based on the idea that lights are not core elements of a transportation facility or that there are non-transportation sources that can be used.

7.6.2 Options for Funding

Capital funding for pedestrian-oriented lighting projects is available from a broad range of sources, including conventional transportation grants, non-transportation grants, assessment districts, development agreements, and public-private partnerships.

Conventional transportation grants are available from federal, state, regional, and local sources. The potential sources and the mechanisms for obtaining them are highly localized and change frequently. In many urban locations, a county or regional agency may prioritize federal or state funding among localities. Therefore, funding options need to be assessed at the local level.

The federal Highway Safety Improvement Program (HSIP) is a primary transportation source, typically administered by the states. Competitiveness focuses especially on addressing safety issues.

Non-transportation grant sources include: redevelopment districts, streetscape grants, economic development grants, and energy innovations funding. Recently, the Energy Efficiency and Conservation Block Grant Program, administered by the federal Departments of Energy and Commerce, has been especially useful for LED retrofits. At the state level, energy conservation grants were offered by the New York Power Authority as upfront funding for LED conversions that municipalities could pay back with energy savings.[31] Consulting support and procurement discounts were also provided. The aim of the 2018 program was to spur the conversion of 500,000 street lights statewide.

Redevelopment district funding typically is used for "blighted" areas or those ostensibly in need of economic rejuvenation. Bonds are floated by a redevelopment agency to pay for infrastructure improvements and sometimes building demolition. Bonds are repaid with tax increments (additional property taxes generated by the higher property values enabled by the new infrastructure and new development).

Other economic development funding is another prime source. For instance, Philadelphia's 'Chestnut Hill Business District received a $225,000 grant from Pennsylvania's Department of Community Development and Economic Development Multimodal Transportation Fund for Pedestrian Lighting toward the $4.5 million costs of replacing 188 pedestrian light fixtures.[32]

Technological change in the field can be leveraged for *innovations funding*. One manufacturer received a federal Department of Energy grant to develop a concept for an outdoor Organic Light Emitting Diode (OLED) luminaire using solar energy for lighting pedestrian areas.[33] (OLEDs have the potential advantage over conventional LEDs of being smaller, thinner, more flexible, and can even be used as pixels in televisions.)

Assessment districts are often organized by local governments to obtain funding from properties, and the city or county may also directly contribute funding. Funding is primarily obtained by assessing businesses and residences based on their proportional share of street frontage or some other formula determining benefits and possibly ability to pay. Assessment districts typically require the approval of a majority of businesses or residents (or those with the majority of property value). For example, a recent Santa Monica assessment district for new street lights found that the cost estimate for a typical residential parcel with 7,500 square feet of land was roughly $6,250.[34]

Development agreements can require the developer to install and possibly maintain roadway lighting and/or pedestrian-scale lighting. Development agreements are legal documents between the developer and local government specifying the government's grant of real estate development rights and guarantees of expedited permit consideration, often in exchange for financial or in-kind contributions from the developer. In San Francisco, California Pacific Medical Center (CPMC) a division of Sutter Health, provided approximately $80 million in total funding for neighborhood and community needs in the development agreement for two new hospitals and other medical center development. This included a set-aside of $4.25 million for 97 new street lights in the low-income Tenderloin neighborhood, plus pedestrian-scale lighting near an existing campus in the more affluent Pacific Heights neighborhood.[35]

Public-private partnerships (PPPs) are especially useful for retrofits providing LED lighting. The locality can enter into a contract with a private party to undertake a street lighting modification. The private business generally pays upfront capital costs, but then is reimbursed by the local government with periodic payments supported by savings from reduced electricity costs (and possibly reduced maintenance).

Smart lighting sensor data provide additional opportunities for obtaining private or public funding. Traffic volume and congestion data are valuable to online driving apps, and air quality data could have monetary value for certain agencies. One University of Michigan professor has suggested that data sales may become a major source of public infrastructure finance.[36]

Operations and maintenance funding is available from a range of sources. The City of Los Angeles street lighting operations and maintenance are funded from an assessment collected primarily through property taxes, as is common in the US. Business Improvement Districts may fund operations/main-

Detroit Public-Private Partnership

Detroit pioneered the public-private approach to funding LED retrofits in 2014.[37] Although this arrangement was used for freeway, bridge, and tunnel lighting, the principles should be applicable to other types of lighting. The Michigan Department of Transportation (MDOT) entered into a PPP agreement to retrofit the Detroit metro region freeway lighting with a consortium that included financial partners, a design-build contractor, an engineering designer, and an operations/maintenance provider. The proposing team was selected based on technical qualifications, experience, and legal/financial capability.

The contract term was 15 years, with LED installation in the first two years. MDOT made two payments during construction, then started quarterly payments for the 13-year operating period. Quarterly service payments were based on a formula that included:

o Energy savings
o Costs to design, construct, finance, operate, and maintain the system
o Inflation
o Penalties for non-compliance events (lack of lighting) and for maintenance-related lane closures

Ownership of the new lighting infrastructure reverts from the consortium to MDOT after 15 years. Economists prepared a Strengths/Weaknesses/Opportunities/Threats (SWOT) matrix evaluating the partnership. They found the arrangement, especially the risk sharing, likely to be a "win-win" situation, better for the public sector than direct provision of enhanced lighting.

tenance and installation costs of new lighting through assessments, typically also collected through property taxes. After LED conversions, public-private partnerships may reimburse street lighting operational costs to the contractor or consortium as part of a regular payment that also includes capital cost reimbursement. Smart city applications offer the opportunity to collect fees from agencies and organizations that "rent" smart pole space, power, and possibly transmitters for sensors and other devices.

7.7 Institutional Issues

Although responsibilities for street and pedestrian light installation and operations/maintenance vary by location, these are frequently split among multiple parties. Often these include both public and private organizations. The fragmentation of responsibilities can complicate initiatives to improve pedestrian safety and walkability.

7.7.1 Governmental Organization

Street lighting is generally considered a core municipal government function. However, responsibilities vary greatly by locality. Often responsibilities for street lighting, pedestrian safety, and other pedestrian facilities (like sidewalks) are split among multiple agencies. As in San Francisco, there may be separate agencies that design and operate street lighting, plus several different agencies involved in transportation planning, design, and funding. Public health agencies are increasingly interested in transportation safety and urban quality of life. Law enforcement, prosecutors, and public defenders may be involved in lighting related to pedestrian injury cases.

Regional and state agencies often work closely with local governments on such issues. They may be important funders of local lighting improvements.

7.7.2 Non-Governmental Stakeholders

There are numerous non-governmental stakeholders who are or should be involved in planning and design decisions regarding lighting. These include utilities, business owners and associations, neighborhood associations, community groups interested in safety or crime, and astronomical observatories.

Some non-governmental organizations are keenly interested in lighting issues and ready to partner with governments and utilities. These include business associations and astronomical observatories.

▶ Figure 7.5

Dimmed Street Lights in Southwest Detroit. The Detroit Public Lighting Authority in a federal lawsuit against the contractor stated that upwards of 20,000 LED lights were "prematurely burning and dimming out." Courtesy of The Detroit News, Clarence Tabb, Jr., photographer.

For example, the Lick Observatory worked closely with the City of San Jose and other adjacent cities to reduce light pollution.[38] The county government of the Big Island of Hawaii, and in Arizona, the Cities of Tucson and Flagstaff plus Pima County have adopted lighting ordinances recognizing the needs of nearby observatories.

The Southwest Detroit Business Association raised over $6 million from government grants and foundations for about 170 new street lights for a 2.2-mile segment of Vernor Highway.[39] This responded to a severe long-term outage situation where over half of Detroit's street lights were not working. (See Figure 7.5.) Installation was managed by the Michigan Department of Transportation. The project was completed in 2019 as part of a major streetscape improvement project, and according to the Business Association website: "Business owners and residents jumped for joy and kissed light poles when the lights came on."[40]

7.7.3 Improving Inter-Organizational Coordination

Improving coordination among different organizations to enhance lighting revolves around three key components:

1. Share information
2. Work out coordination agreements before a critical need
3. Leverage strengths of each organization

Information sharing requires communication about each organization or agency's mission, goals, resources, and constraints. One advantage of a citywide pedestrian lighting plan is that it provides the impetus for organizations to meet and exchange information and ideas. However, even without a major planning effort, information sharing can start with limited initiatives, such as a brown bag lunch to discuss lighting concerns and funding.

Developing coordination agreements even before a plan or project is initiated can be worthwhile. For example, agreements can be developed about requirements to install pedestrian-scale lighting and who will pay for capital costs and increased electricity and maintenance costs. If there is a pedestrian fatality review team (similar to the public health concept of the fetal and infant mortality review team), composed of multiple agencies, that could by agreement include a lighting condition review for after-dark fatalities.

Leveraging organizational strengths should extend beyond the formal mission of each organization to take advantage of unique staff talents and interests. For example, one agency may be particularly experienced at obtaining funding, another at community involvement, and a third at the technical requirements of lighting design.

Notes

1. Illuminating Engineering Society (IES), *RP 8-18: Recommended Practice for Lighting Roadway and Parking Facilities* (New York: IES, 2018), 5–6.

2. Paul Lutkevich, Don McLean, and Joseph Cheung, *FHWA Roadway Lighting Handbook* (Washington, DC: FHWA, 2012); Don McLean et al. "Appendix B-1: Warrants for Lighting Arterial, Collector and Local Roads," *Guide for the Design of Roadway Lighting, Volume 2* (Ottawa, ON: Transportation Association of Canada, 2006).

3. American Association of State Highway and Transportation Officials (AASHTO), *Highway Safety Manual* (Washington, DC: AASHTO, 2010), http://www. highwaysafetymanual.org/Pages/default.aspx.

4. Seattle Department of Transportation, *Pedestrian Lighting Citywide Plan* (Seattle: SDOT, 2012).

5. San Francisco Planning Department, "WalkFirst Phase 2: Pedestrian Safety Partnership," 2014, https://sfgov.org/sfplanningarchive/walkfirst-phase-2-pedestrian-safety-prioritization.

6. Clanton & Associates for City of San Jose, *Public Streetlight Design Guide* (San Jose: City of San Jose, 2016), http://www.sanjoseca.gov/DocumentCenter/Home/View/242.

7. Auerbach-Glasow for the City of San Jose Redevelopment Agency, *San Jose Downtown Street and Pedestrian Lighting Master Plan* (San Jose: SJ Redevelopment Agency, 2017), 15. https://www.sanjoseca.gov/home/showdocument?id=32537.

8. SDOT, *Pedestrian Lighting Citywide Plan*, 6.

9. Seattle Department of Transportation, *City of Seattle Pedestrian Master Plan* (Seattle, Washington: SDOT, 2017), http://www.seattle.gov/transportation/document-library/citywide-plans/modal-plans/pedestrian-master-plan.

10. Seattle Department of Transportation, *Streets Illustrated: Right-of-Way Improvements Manual* (Seattle, Washington: SDOT, 2019), https://streetsillustrated.seattle.gov/overview/vision-for-seattles-new-streets/.

11. San Francisco Public Utilities Commission, "Streetlight Catalogue," 2020, https://sfwater.org/index.aspx?page=920; San Francisco Public Utilities Commission, "Exceptions to the Streetlight Catalogue," https://sfwater.org/modules/showdocument.aspx?documentid=4896.

12. Charles V. Zegeer et al. for Federal Highway Administration, *How to Develop a Pedestrian Safety Action Plan: Final Report* (Chapel Hill: University of North Carolina Highway Safety Research Center, 2006), https://safety.fhwa.dot.gov/ped_bike/ped_focus/docs/fhwasa0512.pdf.

13. San Francisco Planning Dept., *WalkFirst Phase 2*.

14. Debbie Herridge, "What Are SMART Objectives and How Do I Apply Them?" Professional Academy website, accessed April 17, 2021, https://www.professionalacademy.com/blogs-and-advice/what-are-smart-objectives-and-how-do-i-apply-them.

15. Charles V. Zegeer et al. for FHWA, "PEDSAFE: Pedestrian Safety Guide and Countermeasure Selection System," Ped/BikeSafe website, 2013, http://www.pedbikesafe.org/pedsafe/.

16. Jacobs Engineering and San Francisco Municipal Transportation Agency (SFMTA), *Balboa Park Station Capacity and Conceptual Engineering Study: Final Report* (San Francisco: SFMTA, 2012).

17. Jacobs Engineering and SFMTA, *Balboa Park Study*, 53.

18. "Primer in CPTED – What Is CPTED?" Crime Prevention through Environmental Design website, accessed April 4, 2021, https://www.cpted.net/Primer-in-CPTED.

19. City of Portland Bureau of Planning and Sustainability, *The Portland Plan Progress Report* (Portland, Oregon: City of Portland, 2017), https://www.portlandonline.com/portlandplan/index.cfm?a=632343&c=45722.

20. Dave Davis et al., *Participation Tools for Better Community Planning*, 2nd edition (Sacramento, CA: Local Government Commission, 2013), https://lgc.org/wordpress/docs/freepub/community_design/guides/Participation_Tools_for_Better_Community_Planning.pdf.

21. UCLA Center for Occupational and Environmental Health, "Walkability and Pedestrian Safety in Boyle Heights Using the Pedestrian Environmental Quality Index (PEQI)," National Association of City Transportation Officials website, 2013, https://nacto.org/wp-content/uploads/2015/04/Pedestrian-Environmental-Quality-Index-Part-I.pdf.

22. SFMTA, *Western Addition Community Based Transportation Plan: Community Outreach: What Did the Western Addition Community Say?* (San Francisco: SFMTA, 2017), https://www.sfmta.com/sites/default/files/projects/2017/Western%20Addition%20CBTP_Adopted_Reduced_Part2_Community%20%20Outreach.pdf.

23. City of Boston Mayor's Office of New Urban Mechanics, "Beta Blocks: Exploring New Approaches for Community-Led Innovation in Public Spaces," City of Boston website, November 4, 2020, https://www.boston.gov/departments/new-urban-mechanics/beta-blocks.

24. Plannng4Places and International Nighttime Design Initiative, *Municipal Smart City Streetlight Conversion and Evolving Technology Guidebook* (Albany: New York State Capital District Capital Transportation Committee, 2020), https://www.researchgate.net/publication/343376085_Municipal_Smart_City_Street_Light_Conversion_Evolving_Technology_Guidebook.

25. Kenneth Appleman, Ute Beseneck, and Leni Schwendinger, "Deployment of Responsive Lighting by the International Nighttime Design Initiative (NTD)," ResearchGate website, January 21, 2019, https://www.researchgate.net/publication/330510900_Deployment_of_Responsive_Lighting_by_International_Nighttime_Design_Initiative.

26. University of Utah Occupational and Environmental Health & Safety, "Walk after Dark," University of Utah website, 2018, https://attheu.utah.edu/tag/walk-after-dark/.

27. U. of Utah website, https://attheu.utah.edu/tag/walk-after-dark/.

28. Johan Rahm and Maria Johansson, "Assessing the Pedestrian Response to Urban Outdoor Lighting: A Full-Scale Laboratory Study," *PLOS One*, October 4, 2018, https://journals.plos.org/plosone/article?id=10.1371/journal.pone.0204638.

29. Smart Outdoor Lighting Alliance, "Davis, CA LED Streetlight Retrofit," 2016, http://volt.org/lessons-learned-davis-ca-led-streetlight-retrofit/.

30. Sean Doyle, "Our Transportation System Values Some Lives More than Others," Smart Growth America website, November 6, 2019, https://smartgrowthamerica.org/our-transportation-system-values-some-lives-more-than-others/.

31. "Governor Cuomo Announces Smart Street Lighting New Program for All Municipalities across the State," New York State Governor's Office website, February 19, 2018, https://www.governor.ny.gov/news/governor-cuomo-announces-smart-street-lighting-ny-program-all-municipalities-across-state.

32. Sue Ann Rybak, "CHBD Receives $225,000 Grant for Pedestrian Lighting Project," *Chestnut Hill Local*, June 12, 2019, https://www.chestnuthilllocal.com/2019/06/12/chbd-receives-225000-grant-for-pedestrian-lighting-project/.

33. US Office of Energy Efficiency and Renewable Energy, "Outdoor OLED Luminaire Using Solar Energy for Lighting Pedestrian Areas," US Department of Energy website, 2015, https://www.energy.gov/eere/buildings/downloads/outdoor-oled-luminaire-using-solar-energy-lighting-pedestrian-areas.

34. Santa Monica Public Works Department, "Streetlight Installation," City of Santa Monica website, 2019, https://www.smgov.net/Departments/PublicWorks/ContentCivEng.aspx?id=8621.

35. Brock Keeling, "97 New Streetlights Illuminate the Tenderloin," *Curbed San Francisco* website, January 3, 2019, https://sf.curbed.com/2019/1/3/18167159/new-streetlights-lights-lamps-tenderloin-sf-crime-safety.

36. Peter Adriaens, "There's Another Way to Pay for Infrastructure Projects," *Bloomberg CityLab*, April 7, 2021, https://www.bloomberg.com/news/articles/2021-04-07/use-data-not-taxes-to-pay-for-infrastructure.

37. Rui Cunha Marques and R. Richard Geddes, "The Use of PPP Arrangements in Street Lighting: A Win-Win Option?" *Annals of Public and Cooperation Economics* 90, no. 2 (2019): 311–327, https://onlinelibrary.wiley.com/doi/epdf/10.1111/apce.12229.

38. San Jose City Council, "Outdoor Lighting on Private Developments," March 1, 1983.

39. David Muller, "Contractor for Detroit Neighborhood's Independent Streetlight Project to be Named Next Month," Michigan Live website, Updated April 3, 2019, https://www.mlive.com/business/detroit/2013/10/contractor_for_detroit_neighbo.html.

40. Southwest Detroit Business Association, "What is the SDBA Streetscape Project?," SDBA website, 2019, https://southwestdetroit.com/community/streetscape-project/.

Chapter 8

Integrating Pedestrian Lighting into Transportation Design, Operations, and Maintenance

8.1 Purpose and Scope of This Chapter

The previous chapter covered the consideration of lighting for pedestrians in planning and policy analysis. This chapter addresses the next phases in project development (design, operations, and maintenance), focusing on street and walkway projects.

Lighting design is routinely undertaken when new streets are built, often as part of land development. A separated bike/pedestrian path or trail may require lighting if it will be open after dark. Lighting upgrades are also an option with minor upgrades of roadways.

This chapter first discusses the broad principles of lighting that should be considered in the transportation facility design process, as well as the process followed by roadway lighting experts who are part of the design team. It then discusses specific design issues for various facilities in the street right-of-way used by pedestrians (intersections, pedestrian crossings, bus stops, and round-abouts). Walkway design (both in the street right-of-way and for separated facilities) is examined next. (Parking facilities are not included, but they are addressed by the Illuminating Engineering Society's *Recommended Practice RP-8-18*.) Road safety audits are another opportunity often overlooked for ana-lyzing and improving existing or planned lighting. Finally, the chapter considers other transportation operations and maintenance tasks (including responding to citizen complaints and minor upgrades of facilities).

While this chapter briefly addresses placemaking and aesthetic issues considered in transportation design, Chapter 9 provides a more extensive treat-ment of this topic. Chapter 10 then includes considerations of how technologi-cal advances and social changes may affect the design process itself.

8.2 Design Principles and Considerations

The lighting design process and transportation engineering design process share important features and challenges. They start with a consideration of the needs and characteristics of users and the project setting. They involve consid-eration of a wide range of similar factors, often involving difficult trade-off deci-

DOI: 10.4324/9781003149750-8

sions. Work by different specialists must be closely coordinated. Quality control and quality assurance steps are undertaken to minimize errors or omissions.

8.2.1 Pedestrian Lighting Needs

Pedestrian lighting needs must be taken into consideration during the design or redesign of transportation facilities. The capabilities of a range of pedestrians, including those with mobility or visual challenges, need to be considered. The pedestrian's needs after dark include:

- Ability of pedestrians to see details of other pedestrians, signs, and building facades in order to feel secure and orient themselves
- Ability of pedestrians to avoid hazards and obstructions
- Visibility of pedestrians to drivers and bicyclists
- Desire of pedestrians to have a sense of place and experience a visually pleasing environment

To feel **secure and well-oriented**, pedestrians need to be able to recognize faces and see the actions of other pedestrians. They need to be able to read signs, identify building entrances, and make other key observations from a convenient distance.

Pedestrians need to **avoid hazards** while crossing streets or walking along a sidewalk or separated walkway. Potholes, loose objects, tree roots, uneven sidewalk joints, bicyclists on sidewalks, and low-hanging building projections are examples of possible hazards. For this purpose, lighting needs to illuminate the pavement level and up to approximately seven feet above. Contrast and color rendition should be sufficient so the pedestrian can size up hazards early enough to avoid them. Tree canopies or other objects can interfere with needed illumination. Pedestrians with mobility or vision limitations may have different visual needs, such as a heightened need to perceive slipping hazards.

Ensuring **visibility of pedestrians** to drivers and bicyclists for safety requires adequate illumination as pedestrians approach and undertake street crossings. Foliage or other objects near crossings may block the illumination of pedestrians and clear views of them by drivers. Glare or unnecessary distractions to drivers can also be harmful. (Glare sources may be difficult to control as glare may come from auto headlights or privately owned lighting.) The contrast between the pedestrian and the background is also important. Pavement reflec-

Pedestrians need to see and be seen: to see obstacles, threats, and features of their environment and to be seen by drivers and bicyclists. These needs should drive the lighting design process.

tivity, the spectral characteristics of the lighting, and pedestrian clothing also affect visibility.

To fully **appreciate walking** and their environment, pedestrians benefit from a sense of place and a visually interesting pedestrian realm. The quality of the illumination and the lighting equipment itself are important factors in place-making, as discussed in more detail in Chapter 9. An example of integrating lighting into a transportation design for a South Bend, Indiana complete streets project is described in the sidebar.

Accessibility considerations are critical. Designing walkways to be fully accessible provides the "universal design" benefits of helping both those with mobility disabilities and those with other challenges, such as the elderly, pregnant women, small children, and those recovering from surgery or injuries.

South Bend Smart Streets Project

The City of South Bend, Indiana received a Complete Streets Initiative award from the National Complete Streets Coalition for conversion of multiple downtown one-way streets to more pedestrian-friendly two-way streets.[1] (See Figure 8.1.) The Main Street/Michigan Street/Martin Luther King Boulevard project in downtown South Bend was a key component.[2]

The existing one-way streets raised concerns that they provided a poor pedestrian environment, facilitating higher speeds, so the conversion intended to improve safety and efficiency by reestablishing two-way traffic patterns and better accommodating pedestrians, cyclists, and transit riders. The City aimed to improve pedestrian safety and sense of place. The project also sought to increase property values, business revenue, and the number of people walking and bicycling while decreasing crime.

Lighting was a major project focus to improve safety and security. Cobrahead style lighting on tall poles was replaced with decorative, pedestrian-scale lighting. Different poles and luminaires were used according to the setting. Pedestrian lights were installed on 16-foot poles, with single or double ornamental luminaires as necessary to create the desired lighting level. Poles were spaced approximately every 80 feet. Taller poles, up to 23 feet, were used at the two roundabouts near the north end of the project. Lighted bollards (three feet high) were placed in the protected bike lane/shared path along MLK Boulevard at street intersections. Responding to public concerns over high pedestrian volumes crossing at the two roundabouts near the city's largest hospital, in-pavement lights were installed at roundabout crosswalks.

Standard light-emitting diodes (LED) lights were used throughout the project. The goal of the lighting design was to achieve approximately 1.0 foot-candle illumination along the sidewalks while limiting light spillover into the second floors of adjacent buildings.

After the implementation of this project, more than $100 million in investment was made downtown, significantly higher than in preceding years. Downtown saw more restaurants, hotels, and apartments. Traffic counts demonstrate that the number of pass-through vehicles on the streets has decreased. Pedestrian and bicycle usage, especially along MLK Blvd., increased substantially.

South Bend, Indiana
Complete Streets
Project. Courtesy
of Krista Lawson,
photographer,
and American
Structurepoint.

8.2.2 Design Considerations

Lighting designers start with the project goals and consider especially these factors:

- Safety
- Cost
- Maintenance
- Aesthetics
- Environmental impacts
- Site conditions

Safety involves consideration of the visibility of potential hazards and of pedestrians themselves, while limiting glare or distractions from excessive illumination of unnecessary features (as discussed above). Also, the potential hazards of roadside equipment need to be considered as fixed objects that may be hit by vehicles or impediments to clear sight lines. For this reason, limiting the number of luminaires and poles, while achieving visibility objectives, is important. The particular characteristics of the location need to be considered by, for example, consulting collision data and reports.

    Cost includes the range of life cycle costs (capital and operations/maintenance costs, including replacements). Even if costs will definitely be under the budget, cost-effectiveness needs to be considered. Cost/benefit analysis is useful for larger projects. Costs are also highly variable with location, affected by such factors as convenient access to electrical service.

Maintenance needs will be reflected in costs, but they are important also in their impact on the reliability of lighting service and staffing levels. Designers need to consider such factors as the potential frequency of required maintenance, the difficulty in replacing parts, and the physical difficulty in reaching luminaires. Outages raise safety and appearance issues. Solid-state devices such as LEDs and smart lighting equipment, such as smart nodes and environmental sensors, have very different maintenance needs than traditional lighting equipment.

Aesthetics of lighting equipment includes selecting poles in scale with surroundings and complementing street furniture. In some districts, ornamental poles and luminaires are desired by the municipality or other stakeholders. Color appearance of lighting is also a concern to many stakeholders.

Environmental impacts primarily involve light pollution (glare, skyglow, and light trespass), as well as energy use and greenhouse gas emissions. The designer tries to control these through focusing illumination on the roadway or other transportation facility. Potential strategies include physical shielding, dimming, and placement of luminaires. The light source technology and spectral characteristics also affect these impacts. Special area needs, such as proximity to sensitive wildlife, should be considered.

Site conditions may affect illumination needs or constrain the placement of light poles. Examples include:

- Trees and bushes that block light
- Severe weather conditions that reduce visibility
- Corrosion of lighting equipment due to ocean mists or fog
- Shadows across transportation facilities from large buildings that impair visibility
- Narrow right-of-way that limits placement of light poles
- Driveways
- Utility conflicts with potential foundation locations
- Transition zones where lighting needs and impacts change
- Availability of electrical service
- Proximity to sensitive land uses or special facilities like airports

8.3 Lighting Design Process

The lighting design process involves a set of steps similar to transportation facilities design. (See Figure 8.2, from the IES *Recommended Practice RP-8-18.*) For a roadway or walkway project, lighting design is guided by the overall transportation concept and should be integrated into the transportation and civil design process. Lighting issues and needs should be considered at the outset of the project. Design should also be informed by research and community participation.[3] The interaction of lighting and other design elements also should be addressed. (For example, the impact of additional light poles on effective sidewalk width and roadside hazards needs to be considered.) Quality control/assurance should assess lighting design and products throughout (including for

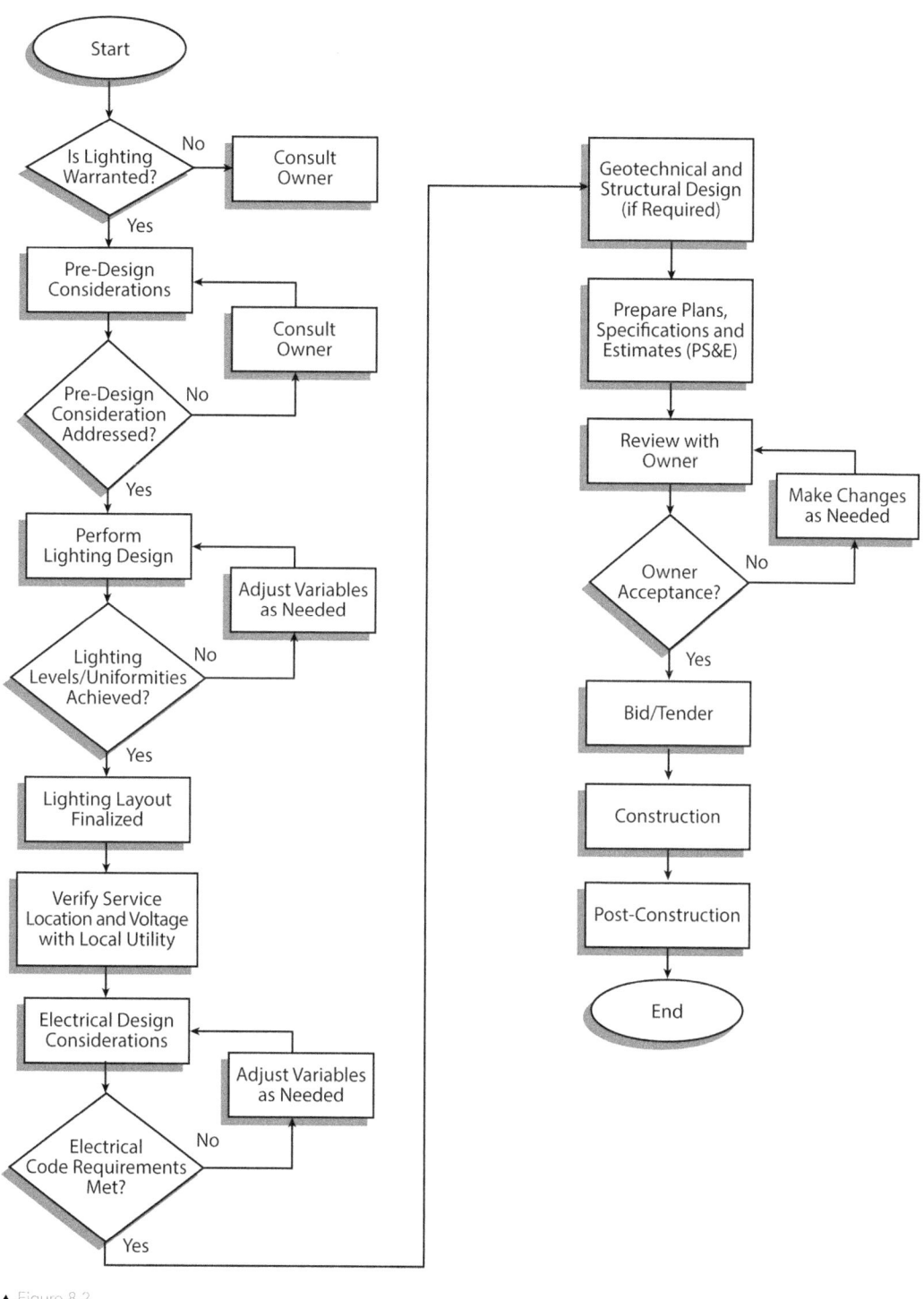

▲ Figure 8.2

Lighting Design Process. Figure reproduced with permission from The Illuminating Engineering Society © ANSI/IES RP-8-18.

construction traffic control). Pilot projects can be extremely helpful in assessing new technologies and obtaining community buy-in, thereby avoiding mass redesign and replacement installation to respond to problems or citizen complaints.

8.3.1 Pre-Design

This phase involves researching the information needed to prepare effective designs, through discussions with the project owner and other stakeholders, as well as extensive document review and site visits. This benefits from a designer looking comprehensively at the project needs and asking effective questions. The designer identifies applicable standards and community goals (including local lighting master plans, illumination standards or criteria), requirements for specific products, procedural requirements, and standard construction detail plans. The designer assesses planned geometrics or layout, utilities, and future use characteristics, considering adjacent development, and physical features. For a street project, the designer can review existing and forecast pedestrian and traffic volumes, collision statistics (and possibly detailed crash reports), and roadway classification. Coordination with urban design and landscape requirements is important. The condition of existing lighting equipment is assessed.

8.3.2 Design

Typically using specialized computer software, such as AGi32, the designer prepares the layout and selects luminaires and poles, using product information on factors such as light source, total light loss factor, pole, and luminaire dimensions.[4] The designer prepares photometric diagrams to show light levels at different locations. Design should take into account key site features such as a tree canopy interfering with light distribution.

Design is typically an iterative activity, and repeated consultation with other designers, the project owner, and stakeholders is needed. As part of design development, mock-ups and renderings can be useful to get project owner and stakeholder approvals. The lighting designer may use specialized electrical and geotechnical design input.

Lighting design needs to be closely integrated into the overall transportation project. This starts with a clear understanding of the goals and constraints of the project and community, as well as the project setting. For example, the designer initially consults any lighting master plans, develops an understanding of the safety and physical characteristics of the setting, and assesses current lighting equipment and needs.

The designer ultimately prepares the Plans, Specifications & Estimates (PS&E) package. The plan drawings typically include plan-view layouts showing existing and planned geometric features (e.g., curbs, sidewalks, crosswalks), utilities, and lighting equipment (pole locations, conduit, and wiring, electrical service location); pole elevation drawings; and electrical diagrams (with service, lighting controls, and branch lighting circuits). Drawings may also include photometric diagrams and plans for temporary construction lighting.

8.3.3 Construction and Post-Construction

The lighting designer is typically involved in the construction phase to review contractor questions, prepare change orders, and help with inspections. The designer also prepares red-line markups of the official plans and records drawings or as-builts, essential for later maintenance or upgrades.

After construction, the lighting designer is involved in integrating and testing different systems. Commissioning includes adjusting the design and explaining the system to those who will be responsible for operations and maintenance.

8.3.4 Quality Assurance and Control (QA/QC)

Concern with quality is essential during all phases of design and construction. Quality assurance focuses more broadly on the process to ensure that quality *requirements* are met in an efficient and transparent manner. Quality control focuses more narrowly on inspecting *products*, such as photometric diagrams and working equipment.

Field measurements with illuminance and luminance meters are the prime method of checking lighting quality after construction. Both design calculations and post-construction measurements should include not just the travel lanes but also the surrounding area, including horizontal and vertical illuminance levels on sidewalks. Ideally, light levels beyond the roadway right-of-way are checked if there are nearby sensitive receptors (like checking light trespass levels at the property lines of adjacent residences).

8.4 Street Design: Intersections and Crossings

The following presents a summary of the special considerations the lighting designer faces for a conventional intersection, roundabout, or mid-block location. Lighting is especially important at these locations not only for safety purposes but also to help pedestrians orient themselves and identify proper crossing points. Attention must be paid to the different physical characteristics of each crossing type and the vehicle flows.

8.4.1 Crossing Types and Lighting Plans

Conventional intersections, roundabouts, and mid-block crosswalks vary in the number of crosswalks and where crosswalks are located relative to vehicle flows. Conventional intersections themselves vary greatly in size, whether there are median islands, what type of traffic control devices are provided, and whether there is on-street parking or driveways close by. Even if crosswalk lines are not painted (in California, many other states, and per the Uniform Vehicle Code) the invisible extension of a sidewalk through an intersection is legally considered a crosswalk, unless explicitly indicated otherwise. In California, pedestrians have the right-of-way in marked and unmarked crosswalks.

8.4.2 Conventional Intersections

Intersections have one of three **types of lighting**:

- Full lighting requires uniform lighting of the entire intersection.
- Partial lighting focuses illumination on key conflict or decision points
- Delineation or beacon lighting merely uses light to notify drivers of the intersection location, without trying to illuminate it effectively

The lighting designer reviews the **criteria or standards** that will shape the design, including:

- Light levels and uniformity requirements
- Lighting zone of the location
- Type of lighting (full vs. partial vs. delineation)
- Special local policies and ordinances
- Any local catalog of approved equipment

Recommended light levels were described in Chapter 2. A key consideration is the difference in illumination in the middle of the intersection versus in the crosswalk versus on the sidewalk corner approaches. IES RP-8-18 calls for "conflict areas" outside the intersection to have illumination 50 percent higher than the street. Pedestrians who are not well illuminated but seem to suddenly appear as they enter the crosswalk, pose a special challenge for drivers. Incorporating surround ratios into the design helps address this issue. As discussed in Chapter 5, the National Academies *Solid-State Lighting Guide* suggests that illumination of shoulders and sidewalks should be at 80 percent of the nearest travel lane. Other influences on pedestrian visibility as they enter crosswalks should be considered, including glare, contrast, pavement reflectivity, spectral characteristics of lighting, as well as foliage or other distractions or visual impediments to drivers.

While glare standards are not provided for intersections, due to the difficulty in measuring this, glare should still be minimized. Minimal high-angle candlepower is helpful in reducing glare.

The *lighting elements* shown in the intersection and crossing lighting plans include:

- Pole placement
- Light source type
- Mounting height
- Pole arm length
- Pole spacing
- Luminaire type and wattage
- Luminaire light output
- Light distribution

Lighting elements need to be coordinated with other roadway features that may constrain the placement of poles and other equipment. *Non-lighting features* that may affect lighting design decisions include:

- Traffic signal equipment (e.g., controller cabinets)
- Pavement type
- Roadway geometrics and right-of-way width
- Bicycle lane/path treatments
- Driveways and property access
- Sidewalks, ramps, bike paths, and street furniture
- Curb ramp locations
- On-street parking treatment
- Signage
- Utilities and electrical conduit
- Bus shelters
- Landscaping
- Street furniture

Traffic engineers and civil engineers are primarily responsible for the design of most of these non-lighting components, with landscape architects or urban designers responsible for landscaping and street furniture. Therefore, coordination among traffic, civil, landscape, and lighting design is critical.

    Pole location is especially important. For example, at signalized intersections, luminaires are often mounted on signal poles to decrease clutter and costs. This may produce sub-optimal vertical illuminance in crosswalks, compared to luminaires mounted on poles in advance of stop bars. Seattle's street improvements manual and Federal Highway Administration (FHWA) guidance state that luminaires should be placed at least 10 feet from the crosswalk to light the side of the pedestrian facing approaching traffic.[5] Poles and luminaires are typically placed on at least two corners, with light distribution so that sidewalk corners and crosswalks meet lighting design standards. At an intersection with only delineation lighting, only one pole is typically provided, on the main road on the approach corner.

    Wider streets may require additional poles and luminaires to provide sufficient crosswalk illumination. At locations with high ambient light levels, higher vertical illuminance (e.g., 30 lux) may be needed.[6]

The desirable clear zone around poles may be shortened due to such constraints as: signal arm lengths, visibility of traffic signal heads, and required access to pedestrian push buttons. Median island light poles are particularly difficult to locate, due to the need to reduce exposure to being hit by turning vehicles, while achieving desired illumination levels.

Pole placement needs to avoid unnecessarily obstructing the pedestrian's path. Any supplemental pedestrian-scale lighting near the intersection may be unnecessary if light levels from roadway lights are sufficient. Vegetation is typically minimized or controlled at intersections to avoid blocking light distribution or driver sightlines to signs, signals, and conflicting vehicles. The lighting designer should coordinate with landscape designers about the size, type, shape, and future growth of trees. While mast arms or longer brackets can help minimize foliage interference with roadway light levels, the adverse impacts on sidewalk light levels from moving the luminaire further away should also be considered. Potential conflicts between tree roots and light poles also should be considered.

The visibility of **traffic control devices** may also be affected by lighting. These essential devices commonly include traffic signals, beacons, STOP or YIELD signs, other signs, crosswalk markings, stop bar markings, and pavement legends. The type of markings and pavement may affect the contrast of the pedestrian against the background, therefore affecting the pedestrian's visibility to drivers.

8.4.3 Roundabouts

Roundabouts generally have **improved safety** compared to standard intersections because of the reduced conflict potential and lower vehicle speeds. The geometrics encourage slower speeds, with vehicles forced by splitter islands to turn from narrow paths into the roundabout circle. Parking is not allowed in or close to the roundabout, reducing obstructions and distractions. (The modern roundabout is different from a "traffic circle," a traffic calming device smaller than the roundabout center island. The traffic circle does not require changes to the curb line at the corners and has much less impact on vehicle paths through the intersection.)

Roundabout lighting treatment differs from that for conventional intersections, in part because the crosswalks are typically set back about 6 to 15 meters (20–50 feet) from the vehicle intersection (circle) through cut-outs in the

Lighting design should focus on the unique lighting needs of pedestrians in or approaching crosswalks. Crosswalk location and characteristics vary significantly among conventional intersections, roundabouts, and mid-block crossings. For example, roundabout crosswalks are typically separated by 6 to 15 meters (20 to 50 feet) from the vehicle circle or intersection.

splitter islands. The visibility of pedestrians is potentially enhanced in modern roundabouts because the driver approaching the crosswalk can focus on pedestrians, without the distraction of identifying conflicting vehicles, which are dealt with after passing the crosswalk. Effective illumination should cover not only the crosswalk itself but also the sidewalk approaches, pavement markings, and signage.

When vehicles approach the roundabout at night, their headlights do not illuminate the potentially conflicting vehicles already in the roundabout, circling from the left. Headlights of the vehicles in the roundabout also only illuminate a short segment ahead.

IES RP-8 Table 12.4 recommends horizontal **illumination levels** for roundabouts that are identical to those for conventional intersections. For the crosswalks, studies generally recommend vertical illuminance levels of 20–40 lux (1.9–3.2 foot-candles) at 1.5 meters (5 feet) high. These levels are significantly higher than recommended horizontal illuminance levels for the intersection outside the crosswalk. Glare (veiling luminance) metrics are not recommended for roundabouts, due to the difficulty of calculating this, but luminaires that meet the appropriate Backlight/Uplight/Glare (BUG) ratings are recommended to reduce glare and light trespass.

Poles should be located on the outside perimeter of the roundabout near each crosswalk approach at a distance from the crosswalk about 0.7 times the mounting height, per RP-8. This may be supplemented by additional poles on the departure side of the crosswalks. Center island lighting is not recommended because light will generally not reach the crosswalks effectively.

A major upgrade of Hillsborough Street in Raleigh, North Carolina, included the conversion of the key Hillsborough and Pullen Road intersection to a roundabout, with lighting improvements integrated into the project. The transportation project, recognized as an outstanding example of context-sensitive transportation design by the Institute of Transportation Engineers, was partly triggered by pedestrian safety concerns.[7] Hillsborough is a key arterial street bordered by North Carolina State University and campus-related commercial uses.

Pedestrian crossings for the roundabout were illuminated with poles at each crosswalk entrance. A combination of 16-foot (5 meter) pedestrian-scale lights and 30-foot (9 meter) roadway lights, with a Correlated Color Temperature (CCT) of 4,000 degree Kelvin, were used, typically spaced about 40–45 feet (12–13 meters) apart.[8] These are being converted to LEDs. The crosswalks on the arterial Hillsborough were offset from the roundabout over 200 feet (61 meters), matching pedestrian desire lines between campus and commercial uses. The crosswalks on the minor street were closer to the roundabout roadway, but still offset roughly 50 feet (15 meters). Decorative fences were used to discourage jaywalking across the roundabout island.

8.4.4 Mid-Block Crossings

Mid-block crossings pose special challenges to transportation and lighting designers because they are less obvious to drivers than intersections, so signs and markings, and possibly crossing signals, are especially important. However, they are geometrically simpler than intersections and typically have little or no cross vehicular traffic. (Adjacent driveways need to be considered.)

High visibility of pedestrians in and approaching the crosswalk is paramount. Positive contrast of the pedestrian against background luminance is important. This typically requires high levels of vertical illumination and mounting poles at least 3 meters (10 feet) before the crosswalk (rather than literally at the crosswalk). Clear zones for poles and breakaway features also need to be considered.

The IES RP-8-18 *Recommended Practice* suggests vertical **illuminance** as measured 1.5 meters (5 feet) high on a grid with points spaced every 0.5 meters as follows:

- 20 lux for locations with low pedestrian conflicts
- 30 lux for locations with medium pedestrian conflicts
- 40 lux for locations with high pedestrian conflicts

Horizontal luminance should also be calculated when the roadway approaching the crosswalk has continuous lighting. Although glare is not calculated, RP-8 recommended luminaire lens parallel to the crosswalk to reduce glare. The 2008 FHWA *Informational Report on Lighting Design for Midblock Crosswalks* highlights the value of white light and higher CCT values in providing better detection of pedestrians and a higher level of facial recognition, borne out by more recent research discussed earlier in this book.[9]

There are numerous **roadway and area features** that affect pole placement and mounting height:

- Roadway geometrics and right-of-way availability
- Shadows from overpasses and other structures
- Sidewalks, bikeways, street furniture, and other physical features
- Signs and signals
- Utilities

Chapter 6 discussed the pros and cons of bollard lighting for midblock crosswalks. It also described supplemental lighting at crosswalks triggered when pedestrians are detected as an experimental treatment. It is important that the baseline light levels meet guidelines and that any special lighting not be distracting.

8.5 Bus Stops

Typically located at intersections or mid-block crossings, bus stops have their own illumination needs. They require illumination sufficient for passengers to walk to and from the stop or shelter safely while staying visible to bus drivers, and ideally to allow passengers to read and feel a sense of security. Bus shelters often have their own lighting. Shelters may be located either selectively at high-volume stops or more broadly based on the criteria of the transit agency or municipality.

San Francisco Solar-Powered Bus Stop Lights and Signs. Courtesy of SFMTA Photo Archive | sfmta.com/photo, 150424_Muni Forward_4' and Jeremy Menzies, SFMTA Photographer.

The Orange County Transportation Authority recommended minimum lighting of 2 footcandles throughout its shelters and 2–5 footcandles "within the bus stop area."[10] Locating stops near existing street lighting may be helpful. Pedestrian-scale lighting should be considered for new lighting.

Santa Monica's Big Blue Bus staff noted that "bus stops that feel unsafe, and with no amenities, result in perceived waiting times that are 2–3 times longer than the actual waiting time."[11] Accordingly, the transit agency added solar lighting and real-time arrival displays to shelters to address this issue.

The San Francisco Municipal Transportation Agency started installing solar-powered lights at its 3,600 bus stops in 2018.[12] The solar-powered lights are intended to make it easier for bus operators to see waiting riders. The lights contain four LEDs under a solar panel/hood using 5 watts, with a glare-reducing hood. (See Figure 8.3.) The lights were prioritized for the higher-frequency rapid lines.

LED lighting at bus stops has significant advantages for energy savings, maintenance, and light quality. Accordingly, Seattle replaced linear lighting at its Westlake Avenue Transit Shelter with LED lighting in 2015.[13]

8.6 Walkway Design

Walkways include sidewalks that are adjacent to the street, separated facilities outside the street right-of-way, and pedestrian tunnels/underpasses or bridges. Walkways may be shared with bicycles, increasing potential conflicts. The frequency and width of driveways or intersections along the walkway also affect the level of conflicts. Separated walkways may be primarily for basic transportation or purely recreational facilities. Recreational facilities generally have more flexibility in design. Walkways are routinely included in parks, plazas, or other special places, and their lighting design may be tightly integrated with other uses in these special environments.

8.6.1 Walkway Types and Lighting Considerations

Both sidewalks in the street right-of-way and those separated walkways (or shared bike/pedestrian paths) open after dark need illumination for pedestrians to view oncoming pedestrians and bicyclists. However, sidewalk lighting is routinely considered as part of the street lighting design.

Sidewalks require higher **light levels** than separated walkways typically due to more frequent conflicts with vehicles at driveways and intersections. However, even paths separated from the street right-of-way may cross driveways or streets, and these conflict areas should be treated as crossings per the previous section. Higher illumination may also be needed at rest areas, benches, curves, signs, or other special features.

Pavement types affect luminance, and therefore visibility of trip and fall hazards. Concrete or asphalt is most typically used for sidewalks, shared-use paths, and urban recreational trails. However, recreational trails may use soils or aggregates, which often have very low reflectance and luminance.

There are numerous **design guides** for various types of walkways and for specialized aspects of walkway design, but the treatment of lighting is often very limited. For example, the National Association of City Transportation Officials (NACTO) *Urban Street Design Guide* covers many elements of sidewalk and street design from a "pedestrian-friendly" perspective, but there is minimal attention to lighting.[14] *Designing Sidewalks and Trails for* Access is an extremely detailed guidebook that provides a detailed roadmap to the design process for pedestrian facilities, although its treatment of lighting is cursory.[15] Specialized design guides address private lighting and light pollution in undeveloped areas with no ambient artificial light (IES LP-11, *Outdoor Environmental Lighting*), security issues (IES G-1-16, *Guideline for Security Lighting for People, Property, and Public Spaces*), and coordination with landscape design (e.g., Janet Lenox Moyer's *The Landscape Lighting Book*).[16]

8.6.2 Sidewalks

Illuminating Engineering Society (IES) *Recommended Practice RP-8-18* and urban street design guides typically call for routine lighting of sidewalks. However, sidewalks are not illuminated in many low-density and suburban locations due to funding considerations and lack of interest by residents. Even in the city of Los Angeles, for example, about 2000 of 6,500 miles of streets do not have street lights.[17]

Lighting for sidewalks may be provided by conventional roadway-scale lights and/or pedestrian-scale lights. The physical components of **lighting equipment** are similar to those for intersections and crossings, as covered in the previous section. Sidewalk lighting needs to be completely coordinated with roadway lighting and to follow appropriate street lighting standards and urban design guidance. IES *Recommended Practice RP-8-18* is a primary reference for sidewalk lighting, including recommended illuminance levels for high, medium, and low pedestrian activity levels (as described in Chapter 2). There are also numerous design guides issued by municipalities that address sidewalk lighting, like San Francisco's *Better Streets Guide* (described in Chapter 7) and New York City's *Street Design Manual*.

There are **challenges** to lighting sidewalks adequately while limiting adverse impacts. Sidewalk lighting effectiveness may be affected by ambient lightings, such as building or plaza lights. Building surfaces that are illuminated can provide a sense of security and reduce shadows, thus improving the visibility of sidewalk users. However, such private illumination may adversely affect residents with light trespass and glare. Privately provided foliage may block light from sidewalks. Light poles are a potential impediment, especially to those with mobility or vision impairments.

8.6.3 Separated Walkways

Lighting for separated walkways is typically provided by lower pedestrian-scale lights, as discussed in Chapter 5. While pedestrian-scale lighting is generally more appropriate to the scale of relatively narrow paths or trails, glare is more difficult to control due to lower mounting heights.

Providing adequate lighting may be **physically constrained** by narrow right-of-way, drainage issues, and lack of existing electrical service points. (The case study in Chapter 6 on the Ballard Bridge discussed how linear lighting on the underside of a bridge was used by the City of Seattle to address these concerns.)

Lighting may be provided **continuously or only at special locations**, such as crossings, rest stops, curves, overlooks, or other special features or conflict points. Lighting that is not continuous may cause difficulties for pedestrians and bicyclists in adapting to varying light levels, so the transitions need to be considered.

Local policies and ordinances that guide design may be issued by the local government or in the case of a separated path or trail, may be provided by

Pedestrian-scale lighting for separated walkways is often constrained by the special characteristics of the setting, while sidewalk lighting is closely related to the lighting of the adjacent roadway.

the facility owner, such as a parks district. For example, the Missoula, Montana parks and trails lighting design standards recommend light levels that can be met by spacing City-standard LED poles and luminaires every 90 to 120 feet.[18] The standards also cover Dark Skies considerations, and they prescribe specific luminaires, poles, conduits, and controls.

The IES has recently issued a **practice guide** covering paths outside the street right-of-way since RP-8-18 does not apply to these facilities. LP-2, *Lighting Practice: Designing Quality Lighting for People in Outdoor Environments* provides higher-level coverage of the topic. (IES was considering issuing additional, more detailed design guidance on this topic as of publication.) LP-2 attempts to balance various lighting needs, with the most basic goals listed first:

1. Orientation and wayfinding
2. Reassurance
3. Hazard safety
4. Atmosphere and enjoyment

In considering guidance on this topic, an IES committee reviewed numerous important factors. For example, illumination levels should be set considering the proximity to particular land uses or physical features and the walking surface. Illumination levels should take into account glare, uplight, potential dimming factors, and short wavelength (bluer light) control. Controlling the spectral content of light sources is challenging because CCT is widely used and simple, but it does not adequately represent the complexity of spectral content.

Buffalo Bayou Trail Lighting, Houston, Texas

The Buffalo Bayou Park includes 20 miles of trails, many near Downtown Houston. Part of that trail network, the Sandy Reed Memorial Trail, includes both a concrete cyclist path and a five-foot-wide asphalt path for runners and walkers. The trail was named one of the top urban trails in the U.S and recognized with an award from the American Society of Landscape Architects.[19] The park showcases improved access to the semi-tropical river, habitat restoration, new landscaping, permanent and temporary public art, and unique lighting.

The non-profit Buffalo Bayou Partnership commissioned a public lighting project with artist Stephen Korns, design firm

L'Observatoire International, and technology development and engineering firm Cooper Perkins.[20] Lighting in a section near downtown Houston is synchronized with phases of the moon, with white or blue supplemental LED lighting depending on the phase of the lunar cycle. During the full moon, trails and bridge structures were planned to be lit with white lighting. During the new moon, blue lighting was planned to reduce sky glow and to support star gazing.[21] (See Figure 8.4.)

The lighting project aimed to establish a visual identity for a neglected corridor, providing a stronger connection to natural features, adding variety and excitement. It started with a *Lighting and Public Art*

Buffalo Bayou Park (Houston, Texas) Pedestrian-Scale Lights with Lunar Cycle Color. Lights under elevated roadway turn blue for new moon. Lighting plan developed by artist Stephen Korns, design firm L'Observatoire International, and technology development and engineering firm Cooper Perkins. Courtesy of Buffalo Bayou Partnership.

Master Plan, which analyzed opportunities and constraints, looking at individual spaces and their relationship to each other and to the rest of downtown Houston.

The *Master Plan* included a range of lighting treatments, supporting a ribbon of light visible from downtown streets. Trail fixtures were mounted on poles and on suitable available supports like bridge columns. These provided dynamic, colored lighting. They were reinforced by the lighting of adjacent trees, bridge supports, and the lawn. The plan proposed a mix of luminaires. For path lighting, metal halide luminaires were planned to be mounted on poles 10–15 feet high, spaced 30 to 45 feet apart, specified at 50–70 watts and 3,000 degrees Kelvin CCT with lunar blue/white LEDs on top.[22] Luminaires were also mounted on bollards spaced 15 to 18 feet apart, with 35 watts power and 3,000 degrees Kelvin CCT. Bridge lighting was metal halide with color filters. More recently, metal halide luminaires have been converted to LED. There are also a variety of luminaires mounted on bridge supports, overhead wires, and handrails. Design elements such as pole placement and light output took into account the varying trail conditions, such as adjacent street lighting, traffic proximity, foliage, and topography. Attention was also paid to minimizing glare, light trespass, and skyglow.

8.6.4 Pedestrian Tunnels and Bridges

Pedestrian tunnels (subways) and bridges are special forms of separated walkways. Pedestrians on such facilities have minimal if any conflicts with motor vehicles, although bicycles may be allowed. Tunnels require daytime lighting and high illuminance levels since there is little or no ambient light. Security is generally a major concern since there are few opportunities for escape from a hostile individual and the pedestrian may be completely out of view of others. (Especially for tunnels, a complete security plan is typically considered, extending beyond lighting to include possible video and audio monitoring, attention to possible hiding places, and possibly regular patrols.)

Due to these security issues, high costs, drainage issues for tunnels, and the reluctance of many pedestrians to use them if they can possibly cross a roadway at grade, new pedestrian bridges are **constructed infrequently** and new tunnels or subways are very rare. Pedestrian bridges are still used to cross rivers, creeks, expressways, freeways, and other major obstacles. They are also used where at-grade walking is hampered by heavy snow or to connect associated buildings.

Pedestrian bridges have high **visibility and visual landmark** potential. Lighting may be considered a prominent aesthetic feature, visible for an extended distance. They also offer unusual mounting opportunities on their sides or overhead structures. Special consideration in the illumination of tunnels is provided for visual adaptation between light levels outside and inside the tunnel.

Spaces under elevated structures in cities also provide opportunities to revitalize overlooked, but important, shared spaces. Effective lighting is an essential component of the design of these spaces.[23] Unlike pedestrian subways or bridges, the pedestrian space often is immediately adjacent to a roadway, so roadway and pedestrian lighting need to be coordinated.

8.7 Road Safety Audits

A road safety audit (RSA) "is a formal safety performance examination of an existing or future road or intersection by an independent, multidisciplinary team. It qualitatively estimates and reports on potential road safety issues and identifies opportunities for improving safety for all road users."[24] The products are a formal RSA report and a written response to its recommendations by a project or facility owner.

An RSA may be used at any **stage of project development**. However, they are most often performed in coordination with design or operations activities.

The New York Department of Transportation found a 20 to 40 percent decrease in crashes at over 300 high-crash locations treated with low-cost improvements recommended by an RSA. Besides the human health **benefits**, there are potential economic benefits:

Road safety audits are typically used with design or operations to identify opportunities for safety improvements. FHWA guidance points broadly to the importance of lighting but emphasizes other components of roadway design.

- Avoiding reconstruction costs to correct safety deficiencies
- Reduced societal costs of fatal and injury collisions
- Reduced liability claims
- Possible reduced maintenance costs

Conducting the RSA and designing its recommended measures typically adds about five percent to engineering design costs.

There are several useful **information resources** for road safety audits, including the FHWA *Road Safety Audit Guidelines* and the FHWA *Pedestrian and Bicyclist Road Safety Audit Guide.*[25] The National Highway Institute has also offered a course on the topic.[26]

The general FHWA *Guidelines* suggest that team members include backgrounds in road safety, traffic operations, and road design. Human factors knowledge and certain specialists may be helpful, although lighting specialists are not explicitly mentioned. Physical inspection in the initial RSA phase and also post-construction should take place in both daylight and darkness. Lighting is to be reviewed in both preliminary design and final design plan review. The *FHWA Guidelines* provide six summary case studies, although none list lighting as a major recommendation.

The **process for conducting the** RSA includes three main phases. (See Figure 8.5.) In the first phase, the project owner and/or design team prepares for the study. In the second phase, a specialized "Road Safety Audit Team" conducts the analysis and presents the findings. In the final phase, the project owner and/or design team responds to the recommendations. The FHWA *Pedestrian and Bicyclist RSA Guide* mentions the importance of field reviews considering the visibility of pedestrians and bicyclists at night, especially at locations where

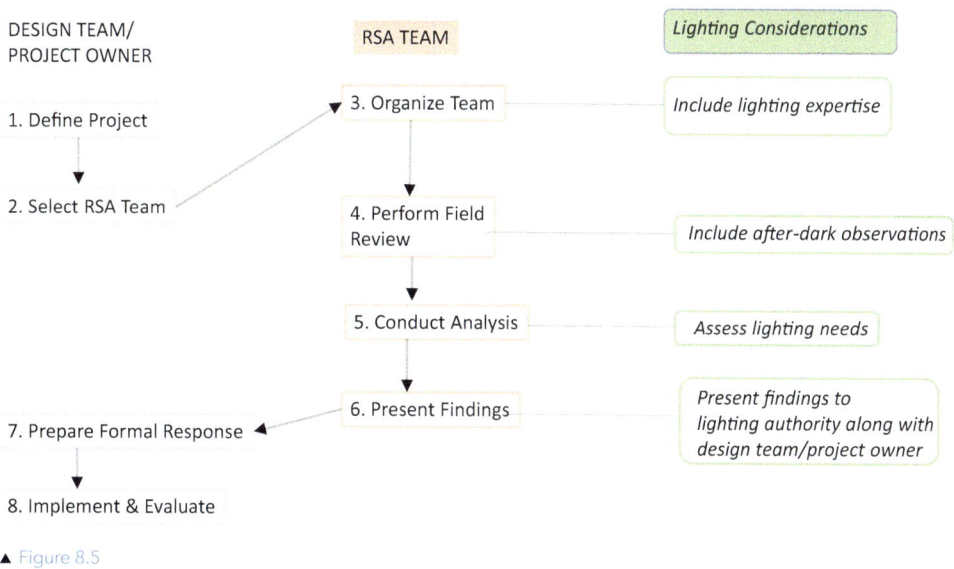

▲ Figure 8.5

Road Safety Audit Process, Responsibilities, and Lighting Considerations.

drivers may be distracted by factors such as conflicting motor vehicle turns. It mentions the need to consider the quality, not just the presence, of lighting, but does not provide any guidance on what is "adequate" lighting.

The Hagerstown/Eastern Panhandle (Maryland) Metropolitan Planning Organization conducted an RSA that addressed lighting and other pedestrian needs on U.S. Highway 40/Cleveland Avenue.[27] This Dual Highway is an important commercial arterial and access to the regional highway network. Data collection included pedestrian crossing counts between 11 PM and midnight at three locations, as well as a questionnaire survey. Recommended improvements included pedestrian-scale lighting at crosswalk entries and bus stop shelters, pedestrian signals, curb ramps, new sidewalk sections, crosswalk warning signs, and a curb bulb-out before a bus turnout. Enhanced enforcement was coordinated with the physical improvements.

8.8 Transportation Operations and Maintenance

Lighting for pedestrians needs to be considered in routine operations and maintenance tasks. These tasks aim to keep facilities operating in good condition, to plan and implement routine upgrades, and to identify and respond to safety problems. These efforts need to consider not just lighting equipment, but other factors such as changes in foliage blocking light.

8.8.1 Operations: Identifying and Responding to Problems

Roadway lighting engineers and managers periodically assess the quality and reliability of their lighting installations. They may follow a prescribed program or respond to public complaints or notices from government colleagues.

Likewise, transportation planners, engineers, and managers assess the safety, efficiency, and reliability of facilities on a routine basis or upon receipt of a public complaint or legal notice. The transportation professional needs to consider to what extent lighting could be a contributor to a facility problem, consulting lighting specialists as needed.

Vehicle crashes disproportionately high after dark compared to daylight crashes at a particular intersection or on a corridor segment are a key indicator of the need to investigate further. Ideally, such locations are identified proactively, such as by an annual review of higher collision locations in the city. Both vehicle/pedestrian crashes and vehicle-only crashes need to be considered. With crashes at a particular location a relatively rare event, a night/day crash rate ratio greater than 1.0 may be influenced by random variations, and not necessarily indicative of a correctible problem. The Transportation Agency for Canada warrants for roadway lighting (described in detail in an FHWA Guide) recommend that a night-to-day crash ratio greater than 2.0 "automatically warrants" lighting installation.[28] As night-to-day crash ratios increase from 1.0 to 2.0, increasing warrant points are added. Guidelines on assessing benefit/cost ratios of lighting improvements based on night/day crash rate ratio and Average Daily Traffic were also provided by a National Cooperative Highway Research Program report.[29]

Public complaints may flag concerns about lighting at particular locations, such as an intersection. However, a citizen may not be able to determine precisely what the problem is. For example, the complaint could be that it is difficult for drivers to see pedestrians at a particular intersection. The most significant problems may be ones that are not identified by a simple check of light levels versus IES recommendations, such as glare from vehicle headlights or private lighting, relatively dark street corners, or an adverse interaction between vehicle headlights and street lights producing poor contrast.

In addition to roadway safety audits, **other types of after-dark inspections** may flag lighting concerns. For example, traffic signs are required to meet retroreflectivity standards, which may be checked by nighttime visual inspection, offering an opportunity to also check lighting conditions.[30]

8.8.2 Operations: Routine Upgrades

Lighting engineers and managers regularly plan upgrades to roadway lighting to decrease costs and improve the quality of lighting. An agency or utility may also make upgrades to reduce malfunctions or to address safety concerns.

Transportation planners, engineers, and managers also routinely plan and implement upgrades of facilities that serve pedestrians, such as installing bus shelters and traffic signals. These efforts also provide an opportunity to assess and upgrade lighting. For example, installing **bus shelters** at high-volume bus stops should trigger an assessment of lighting conditions and needs. Local governments may compile ratings or warrant checks to prioritize intersections for **traffic signals**. The assessment of crash patterns and physical conditions can also consider the potential for lighting improvements.

Typically, a traffic or transportation engineer completes warrant checks to determine whether a signal is justified. (Meeting warrants does not require installation, as engineering judgment and budget constraints need to be considered.) The California traffic signal warrants guidance suggests that, since traffic signals may increase vehicle delay and some types of crashes, alternatives be considered, including roadway lighting "if a disproportionate number of crashes occur at night."[31] Two of the nine warrants are based on factors that would tend also to prioritize the need for lighting: pedestrian volumes and crash experience. The pedestrian volumes warrant is based on possible delays for pedestrians in crossing the major street, based on a combination of pedestrian and motor vehicle volumes. The crash experience warrant also requires that an "adequate trial of alternatives with satisfactory observance and enforcement has failed to reduce crashes." Alternatives could include lighting if after-dark crashes are an issue. In addition, the crash experience warrant requires at least five crashes susceptible to correction by a signal within a 12-month period, plus minimum vehicular or pedestrian volumes.

8.8.3 Maintenance

Street light maintenance includes both preventive programs and responses to problems. Some maintenance is performed directly on lighting equipment, but proper maintenance also involves trees or other objects not controlled by the street lighting department or utility. The key recommendations of IES *Recommended Practice RP-8-18* address:

- Types of maintenance problems
- Preventive maintenance programs
- Potential traffic hazards from lighting maintenance activities
- Maintenance reporting systems

Types of maintenance problems that are especially likely to affect the safety and operations of transportation facilities include luminaire failures, degradation of light levels over time, and obstruction of lights and photocells. Degradation of light output is also due to dirt, so regular cleaning of the lens is desirable.

High-voltage circuits have been a long-standing problem for Los Angeles and other cities. Dilapidated underground wiring has required the majority of Bureau of Street Lighting (LABSL) maintenance efforts, but the Bureau has also converted these 3,000-volt circuits to lower voltages.[32] Recently copper wire and power theft have become an increasing problem for LABSL. The Bureau has a multi-track program to address this, including hardening access to wiring and using cameras at selected locations.

Preventive maintenance programs focus primarily on replacing luminaires based on their rated life or observed malfunctions. Luminaire replacement provides an opportunity to efficiently inspect poles, bases, and fasteners. High-intensity discharge (HID) luminaires should be replaced before the end of their rated life. LED luminaires rarely completely fail, but rated life can be defined as the point when lumen output decreases to 70 percent.[33] (LED luminaire degradation faster than expected was a serious problem for Detroit's LED conversion program.[34]) Luminaires can be replaced on a spot or group basis. The IES *Recommended Practice RP-8-18* suggests routine night patrols and regular physical inspections.

Tree trimming is often needed to prune foliage blocking light beams (or blocking photocells, switching on lights wastefully during daylight). (See Figure 8.6.) Foliage may also impede traffic or pedestrian movements. Foliage may also block drivers' views of pedestrians approaching crosswalks.

Lighting maintenance needs to include preventive maintenance and management systems to ensure quality. Elements besides lighting equipment, such as the need for tree pruning, should be considered.

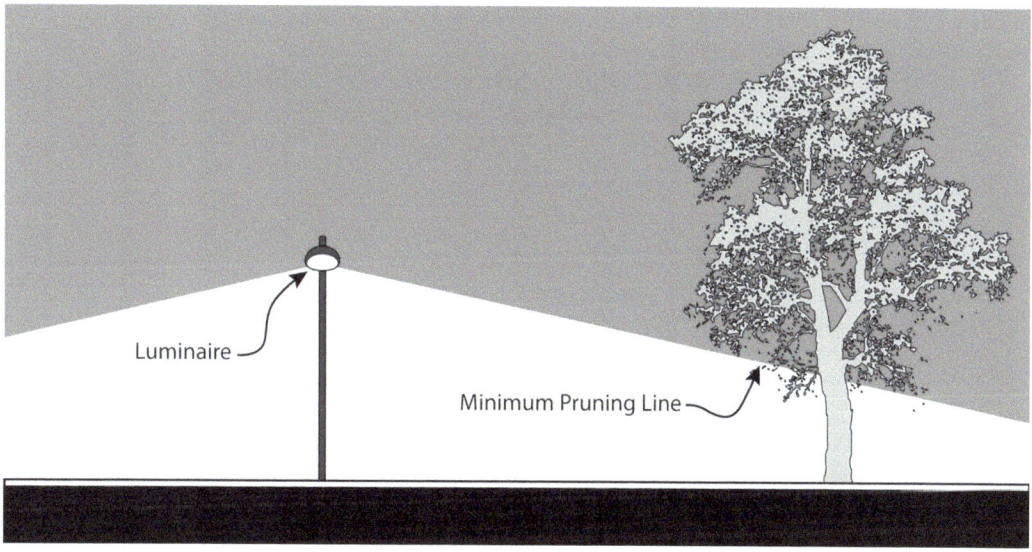

▲ Figure 8.6

Tree Pruning Line. Figure reproduced with permission from The Illuminating Engineering Society © ANSI/IES RP-8-18.

Traffic control is essential if maintenance activities may interfere with traffic or pedestrian movements or impede the visibility of pedestrians. Separate, protected pedestrian paths should be maintained during construction.

Maintenance reporting systems are increasingly automated. (Smart lighting remote monitoring and mapping systems were described in Chapter 6.) IES *Recommended Practice RP 8*-18 also supports standard forms and procedures for recording and respond to complaints from the public, contractors, and police about malfunctions or light levels. Recordkeeping systems that track assets and workflow not only improve response to specific problems, but they also allow for in-depth analysis of maintenance needs, asset inventories, work procedures, and power usage and billing.

Technological advances can be used to improve maintenance in several ways, as demonstrated by the Los Angeles Bureau of Street Lighting (LABSL). The LABSL uses a Geographic Information System (GIS) and a network of smart nodes to map street light conditions and performance along with quick notification of problems and analysis of energy and maintenance trends.[35] The LABSL found that smart nodes (on 37,000 of 223,000 street lights) provided rapid notification of outages, faster than relying on residents to report.[36] Having accurate details on the extent and location of outages allowed for a more efficient response to problems. Information on the energy usage of individual luminaires can also point to needed repairs that can avoid early burnout and energy waste.

The LABSL connected the maintenance "Asset Management System" (GIS) and the 311 public complaint computer system.[37] Public complaints or requests can generate a computerized incident notification to LABSL. After the request has been resolved, that information can automatically be sent back to the 311 system for resident notification.

LABSL plans to install 7,000 additional smart nodes in the 2020–2025 period, a pace constrained by evolving technology and funding.[38] Equitable distribution of smart nodes is considered in prioritizing locations.

Notes

1. Nimotalai Azeez, "Announcing the Best Complete Streets Initiatives of 2017," Smart Growth America website, March 21, 2018, https://smartgrowthamerica.org/announcing-best-complete-streets-initiatives-2017/.

2. Michael Davita, Principal Planner, City of South Bend, Indiana, Email, September 30, 2020.

3. Navaz Davoudian, ed., *Urban Lighting for People: Evidence-Based Lighting Design for the Built Environment* (London: RIBA Publishing, 2019).

4. "The Lighting Industry's Premiere Calculation Tool," Lighting Analysts website, accessed March 16, 2021, https://lightinganalysts.com/software-products/agi32/overview/.

5. City of Seattle, *Streets Illustrated: Right-of-Way Improvement Manual* (Seattle: Seattle Department of Transportation, 2019), Section 3.6, https://streetsillustrated.seattle.gov/design-standards/lighting/; Ronald B. Gibbons et al., *Informational Report on Lighting Design for Midblock Crosswalks*, No. FHWA-HRT-08-053 (McLean, Virginia: Federal Highways Administration, 2008), 12, 15.

6. Gibbons et al., *Mid-Block Crosswalks*, 17.

7. Institute of Transportation Engineers and Congress for the New Urbanism, *Implementing Context-Sensitive Design on Multimodal Thoroughfares: A Practitioner's Handbook* (Washington, DC: ITE, 2017).

8. Dustin Brice, Transportation Department, City of Raleigh, North Carolina, Email, January 29, 2021.

9. Ronald Gibbons et al., *Lighting Design for Midblock Crosswalks*.

10. Kimley Horn for Orange County Transportation Authority, *Bus Stop Safety and Design Guidelines*, March 24, 2004, https://nacto.org/wp-content/uploads/2015/04/bus_stop_safety_design_guidelines_kimley.pdf.

11. City of Santa Monica, "Staff Report 3087: Request for Proposals Award for Bus Stop Lighting and Real-Time Displays," October 9, 2018, http://santamonicacityca.iqm2.com/Citizens/Detail_LegiFile.aspx?Frame=&MeetingID=1151&MediaPosition=&ID=3087&CssClass=.

12. Teresa Hammerl, "SFMTA to Install New Bus Stop Signage, Lighting," *Hoodline*, February 14, 2018, https://hoodline.com/2018/02/sfmta-to-install-new-bus-stop-signage-lighting.

13. Norm Mah, "Lighting the City Up with Energy Efficient LED," Seattle Department of Transportation Blog, February 3, 2015, https://sdotblog.seattle.gov/2015/02/03/lighting-the-city-up-with-energy-efficient-led/.

14. National Association of City Transportation Officials, *Urban Street Design Guide* (Washington, DC: Island Press, 2013), https://nacto.org/publication/urban-street-design-guide/.

15. Julie Kirschbaum et al., *Designing Sidewalks and Trails for Access, Part II: Best Practices Design Guide* (Washington, DC: FHWA, 2001), https://safety.fhwa.dot.gov/intersection/other_topics/fhwasa09027/resources/Designing%20Sidewalks%20and%20Trails%20for%20Access.pdf.

16. Janet Lennox Moyer, *The Landscape Lighting Book*, 3rd edition (Hoboken, New Jersey: Wiley, 2013).

17. Megan Hackney, "LA Smart City Online Streetlighting Conference," Los Angeles, September 10, 2020, https://smartcityla.com/.

18. "Section 3: Park and Trail Lighting Design Standards" City of Missoula, Montana website, March 31, 2017, http://www.ci.missoula.mt.us/DocumentCenter/View/38608/03-31-2017-PART-5-SECTION-3-Only?bidId=.

19. Ashley Biggers, Matcha, "The 10 Best Urban Trails in the United States," Superfeet blog, June 9, 2019, https://www.superfeet.com/en-us/blog/the-10-best-urban-trails-in-the-united-states; American Society of Landscape Architects, "Award of Excellence: Buffalo Bayou Promenade," ASLA website, 2009, https://www.asla.org/2009awards/104.html.

20. Stephen Korns and L'Observatoire International, *Buffalo Bayou Lighting and Public Art Master Plan* (New York: L'Observatoire International, 2001), https://www.houstontx.gov/planhouston/sites/default/files/plans/BuffaloBayouLightingandPublicArtMasterPlan_JW.pdf.

21. Maren Berger and Herve Descottes, L'Observatoire International, Email, February 16, 2021.

22. Trudi Smith, Director of Public Relations and Events, Buffalo Bayou Partnership, Email, February 1, 2021.

23. Ozgur Gungor, "NYC Transforms Underpass into Lively Public Space," Design Trust for Public Space website, June 5, 2018, https://www.designtrust.org/news/nyc-transforms-underpasses-lively-public-space/.

24. Synectics Transportation Consultants et al., *FHWA Road Safety Audit Guidelines*, FHWA-SA-06-06, (Washington, DC: FHWA, 2006), https://safety.fhwa.dot.gov/rsa/guidelines/documents/FHWA_SA_06_06.pdf. Elissa Goughnor et al. (VHB and U. of North Carolina), *Pedestrian and Bicyclist Road Safety Audit Guide and Prompt Lists* (Washington, DC: FHWA Office of Safety, 2020), https://safety.fhwa.dot.gov/ped_bike/tools_solve/docs/fhwasa20042.pdf.

25. Synectics et al., *FHWA Road Safety Audit Guidelines*, 1.

26. "Road Safety Audits," Federal Highway Administration website, accessed March 21, 2021, https://safety.fhwa.dot.gov/rsa/training/.

27. Matt Mullenax, "Dual Highway Pedestrian Road Safety Study," Association of Metropolitan Planning Organizations Conference, Baltimore, Maryland, 2019.

28. Paul Lutkevich, Don McLean, and Joseph Cheung, *FHWA Lighting Handbook* (Washington, DC: FHWA, 2012), 34.

29. Mark S. Rea et al., *Review of the Safety Benefits and Other Effects of Roadway Lighting: Final Report*. NCHRP 5-19 (Washington, DC: Transportation Research Board, 2009).

30. Paul J. Carlson and Matt S. Lupes, *Methods for Maintaining Traffic Sign Retroreflectivity*, Pub. No. FHWA-HRT-08-026 (McLean, VA: FHWA Office of Safety Research and Development, 2007), https://safety.fhwa.dot.gov/roadway_dept/night_visib/policy_guide/fhwahrt08026/chapter3.cfm.

31. Caltrans, *California Manual of Uniform Traffic Control Devices* (Sacramento, CA: Caltrans, 2014), 820, https://dot.ca.gov/-/media/dot-media/programs/safety-programs/documents/ca-mutcd/rev-5/camutcd2014-part4-rev5.pdf.

32. Hackney, "LA Smart City Online Streetlighting Conference."

33. National Academies of Science, Engineering, and Medicine, *Solid-State Roadway Lighting Design Guide: Vol. 1: Guidance* (Washington, DC: The National Academies Press, 2020), 5, https://www.nap.edu/catalog/25678/solid-state-roadway-lighting-design-guide-volume-1-guidance.

34. Christine Ferretti, "Detroit to Get $4 Million in Settlement over Defective LED Streetlights," *The Detroit News*, December 9, 2019, https://www.detroitnews.com/story/news/local/detroit-city/2019/12/10/detroit-settlement-defective-led-streetlights-duggan/4182291002/.

35. Megan Hackney and Nathan Dominguez, "LA Smart City Online Streetlighting Conference," Los Angeles, CA, September 10, 2020, https://smartcityla.com/index.php.

36. Los Angeles Bureau of Street Lighting, *LA Lights Strategic Plan 2020-2025* (Los Angeles: BSL, 2020), 17, http://bsl.lacity.org/strategic_plan.html.

37. Megan Hackney, Executive Officer, Los Angeles Bureau of Street Lighting, Email, March 17, 2021.

38. Hackney Email, 2021.

Chapter 9

Placemaking and Aesthetics

Considerations and Options

9.1 Purpose and Scope of This Chapter

Safety and security are paramount goals that guide planning, policy making, design, and operations/maintenance for both transportation and lighting. However, a refined sense of place, improved aesthetics, and enhanced information can also be important benefits of lighting installations, often with secondary economic benefits.

Lighting can improve the visual experience of the pedestrian environment through the qualities of the light itself and also through the appearance of lighting equipment. The street environment is affected not only by public street lighting but also by adjacent private lighting installed outside of the street right-of-way. In addition to the positive impacts of a more inviting pedestrian environment, there may be adverse impacts of light pollution and distracting overlighting.

This chapter first provides an overview of how lighting can improve the sense of place and the visual quality of the pedestrian environment. It then focuses on different types of special lighting, starting with historic or ornamental light poles and luminaires used in many popular tourist destinations and residential neighborhoods, as well as futuristic lighting for innovation districts and tourist meccas. It next covers the role of artistic lighting (including both static lighting and programmable, dynamic lighting). Then this chapter addresses lighting to enhance the appreciation of the natural environment and to convey information to pedestrians. (There is a great deal of overlap among these categories. For example, artistic lighting often uses futuristic features.)

9.2 Considerations and Process

Lighting has the power to reveal the special qualities and dynamism of the built and natural environment. The close relationship of the quality of light and the built environment has been noted by architects such as Santiago Calatrava (designer of New York's World Trade Center Transportation Hub) and sculptors such as Auguste Rodin (who carved *The Thinker* and *The Gates of Hell*). That Paris has been known as the "City of Light" for centuries suggests the power

> Placemaking efforts offer the opportunity to enhance public spaces with special lighting, increasing after-dark access and the vibrancy of the pedestrian environment.

of lighting to boost the appeal and shape the images of urban places. (The nickname actually originated in the 1667 effort by the Lieutenant-General of the Police to increase lighting as a crime deterrent.[1] However, it has since become associated with some 300 illuminated sites, 33 illuminated bridges, the 20,000 light bulbs on the Eiffel Tower, and special holiday lighting.) The concept of a "city of light" has such power that 70 cities and towns banded together to sponsor the organization "Lighting Urban Community International" (LUCI).[2] These cities use light as a tool for social, cultural, and economic development. LUCI promotes research regarding all forms of public lighting and light festivals through international cooperation.

Using lighting to enhance the visual appeal and identity of urban places requires attention to the special characteristics of the space and how it is used. This benefits from a collaboration among lighting specialists, other design professionals, artists, and stakeholders. They aim to improve the visibility, clarity, and attractiveness of the after-dark environment.

IES *Lighting Practice LP-2 Designing Quality Lighting for People in Outdoor Environments* describes a hierarchy or pyramid of pedestrian needs starting with considering the context for design (such as the lighting zone), addressing the most basic requirements of orientation and wayfinding, then reassurance, pedestrian safety, and atmosphere and enjoyment at its pinnacle.[3] A number of other helpful resources focus on integrating different components of streetscape and urban place design. For example, the Project for Public Spaces website provides a number of resources, including case studies and training programs, with some content on outdoor lighting, ranging from articles on using park lighting to combat gang violence in Los Angeles to more general guidance on lighting principles.[4] Conferences, training, and publications by organizations such as the American Planning Association, the Urban Land Institute, the International Downtown Association, the Mayor's Institute of City Design, and Art Place America (the "creative placemaking" archive) address placemaking. General references include: the National Association of City Transportation Officials *Urban Street Design* Guide (discussed in Chapter 2) and Jeff Speck's *Walkable City: How Downtown Can Save America, One Step at a Time*.[5] More specialized references for specific environments can be helpful. Examples include Janet Lennox Moyer's *The Landscape Lighting Book*, Dave Colangelo's *The Building as Screen: A History, Theory and Practice of Massive Media* (on building façade illumination and projections), Allan Jacobs' *Great Streets*, and William H. White's *The Social Life of Small Urban Spaces*.[6]

Although some classic works on urban placemaking, such as *The Death and Life of Great American Cities* by Jane Jacobs, devote only limited attention to lighting, their broad principles can stimulate thought on how lighting could fit into special places. Jane Jacobs noted the potential security value of "good" street lights, but argued that their value was principally in attracting "eyes on the street."[7] This fit with her emphasis on a vibrant public realm, facilitated by wide sidewalks, traditional neighborhood street life, and diversity of land uses and architecture. Jan Gehl complained of "visually chaotic" public lighting caused partly by different approaches and standards used at different periods.[8] In *Public Spaces and Public Life* Gehl, like Jacobs, emphasized the importance of observing how public spaces are actually used and learning from successful examples of placemaking.[9] Kevin Lynch emphasized the importance of legibility, supporting a clear image of the city by clarifying key paths, edges, districts, junctions, meeting places, and landmarks.[10] After-dark legibility is enhanced by effective lighting.

9.2.1 Factors in Improved Sense of Place and Aesthetics

There are several ways that lighting can positively influence the aesthetics and sense of place of urban streets and spaces:

- Increasing the legibility of buildings and streets
- Imparting vibrancy or dynamism
- Enhancing the attractiveness of the pedestrian environment
- Communicating a hierarchy of places and uses
- Highlighting the unique characteristics of a place.

Legibility involves the visibility and clarity of key buildings, signs, and street features after dark so that pedestrians can orient themselves. It may also involve providing information about the area. For example, the Empire State Building and civic buildings in San Francisco and other cities are often illuminated with color coding for particular messages or events. Special lighting can be used to encourage pedestrians to walk along desired paths toward key destinations.[11]

Vibrancy results from a sense of activity and energy, even drama. Lighting can directly increase the vibrancy of a place by emphasizing high-activity zones and also indirectly increase it, by leading to greater pedestrian volumes. Popular historic districts and other tourist meccas often use ornamental or futuristic lighting to underscore a consistent place theme, which heightens the sense of energy. Changing lighting as the tourist progresses through a district or a temporary show also increases the excitement.

Attractiveness of the lighting for a place is a subjective feature. It is influenced by such features of the lighting as color rendition and illumination levels. It also is influenced by lighting equipment, such as ornamental or historic poles and luminaires. Light poles can be used for attractive banners or planters.

Allan Jacobs pointed to the importance of street light poles and luminaires as street furniture that can contribute to the quality of "Great Streets" in his book of the same name.[12] He identified regular pole spacing as a pleasing linear feature. The book compliments luminaires and poles with ornate designs such as on the Paseo de Gracia in Barcelona, Spain, or simple globes like those on Orange Grove Boulevard, Pasadena, California (without addressing possible light pollution impacts).

A hierarchy of places and uses communicates that some places are more public or important after dark. Of course, a major commercial or ceremonial boulevard with higher pedestrian use in the evening typically should have greater illumination levels than a single-family residential street. This will automatically occur if Illuminating Engineering Society (IES) light level recommendations are followed since the recommendations are based heavily on pedestrian and traffic volumes. Ceremonial streets (e.g., those with important cultural buildings like museums) can have civic importance beyond their level of use, justifying higher illumination levels. For example, Philadelphia's Benjamin Franklin Parkway specially illuminated 20 works of art and the facades of eight landmark buildings starting in 2003 and 2004.[13]

Unique characteristics of a place can be emphasized by lighting to instill a sense of pride, belonging, or even wonder among residents or visitors. Lighting can be used to: accent historic or natural features, emphasize technological advances, create artistic highlights, or enhance a celebratory mood.

9.2.2 Process to Improve Placemaking and Aesthetics

Lighting has great potential to facilitate placemaking. Lighting projects provide an opportunity to attract a broad range of stakeholders and experts to participate in the planning and design process. Such projects can bridge public art and infrastructure.

An effective process for making "great urban places," according to the Project for Public Spaces, should be:

- Community-driven
- Visionary
- Functional
- Adaptable
- Inclusive
- Focused on creating destinations
- Context-specific
- Dynamic
- Trans-disciplinary
- Transformative
- Flexible
- Collaborative
- Social.[14]

A primary means of assuring that placemaking and aesthetics are considered in lighting and transportation projects is to include artists, diverse community representatives, and cultural organizations throughout the process. The National Endowment for the Arts "Our Town" program, which has funded lighting projects, emphasizes projects "that integrate arts, culture, and design activities into efforts that strengthen communities by advancing local economic, physical, and/or social outcomes."[15] Grant requirements include partnerships among cultural organizations, local government, and such infrastructure sectors as environment, energy, and transportation. Another important early step is to analyze the area's unique characteristics, history, and goals as they may affect lighting. Visualization exercises can demonstrate the aesthetic value of alternative lighting concepts. Finally, lighting equipment and illumination impacts on the visual quality of the pedestrian environment should be considered before lighting concepts are adopted by policy bodies.

"Smart Everyday Nighttime Design", a research project based in Cartagena, Colombia, a UNESCO- designated World Heritage Site, sought to explore the possibility of building stronger community connections with light.[16] As the project evolved over two years, the outcome, a pilot, was a mix of local creativity with higher-quality lighting. The effort was led by artist/designer Leni Schwendinger for a major engineering firm (Arup) and an interdisciplinary team. Residents and stakeholders contributed to the design of modern LED lanterns that showcased the identity of the historic neighborhood.

9.3 Historic or Ornamental Street Lighting

Historic or ornamental street lighting is used extensively in older residential and commercial districts. Ornamental lighting, such as the common acorn luminaire, sometimes with decorative finials or other touches, is popular with many pedestrians. In some cases, such lighting is a major tourist draw to older commercial districts.

9.3.1 Historic Districts as Tourist Attractions

San Diego, California, Vancouver, British Columbia, and Boston, Massachusetts have Victorian-era-themed decorative lights in historic districts that are major international tourist attractions. The lighting complements and accents historic buildings and infrastructure.

San Diego's Gaslamp Quarter (or "Gaslamp District") is a 16-block area with 94 historic buildings, plus extensive entertainment and nightlife options.[17] It often hosts special events and festivals such as "Taste of the Gaslamp" and "Mardi Gras in the Gaslamp." The ornamental lights are no longer actually gas lamps, with limited exceptions. Electric arc lighting is expected to be converted to LED soon. (See Figure 9.1.)

Several years ago four actual gas-powered lamps were installed in the center of the district. [18] The City and the business association plan to convert 5th Avenue to a pedestrian promenade for 12 hours daily. Lighting is expected to be lowered from typical 18-foot mountings to 12 feet high, more consistently spaced, along with an LED conversion.

Boston's Beacon Hill is the oldest historic district in Massachusetts. There in 1977 Boston Gas installed 16 quaint gas lamps to make Temple Street (with its brick-paved sidewalks) appear more historic. [19] They replaced mercury vapor electric bulbs. In late 2020, the City of Boston started to consider replacing the lamps with LED facsimiles.

◄ Figure 9.1

San Diego
Gaslamp Quarter
with Historic-
Theme Lighting.
© Joanne DiBona
Photography.

Gastown District in Vancouver, British Columbia is actually named after its Victorian-era founder, sea captain, and saloon owner "Gassy Jack" Deighton. The 12-block area is similar to the San Diego historic district, but with higher residential density. In 2012, it was named the fourth-most stylish neighborhood in the world.[20]

There are numerous examples of lighting used to promote the visual appeal and interest of areas by referring to local history. For example, San Francisco's "Main Street," Market Street, uses sculptural "Path of Gold" lamp posts that illustrate the city's Gold Rush roots. These street lights have mini-sculptures at the base showing Gold Rush figures. (See Figure 9.2.) The visual benefits of their illumination creating a "brilliant linear pathway" visible from hilltops over a mile away were noted in a major project Environmental Impact Report.[21] In San Mateo, California, special lamp posts denote locations on the historic El Camino Real, "Royal Road," trod by the founders of the Catholic missions that were instrumental in the early growth of the state. The highly ornate lamp posts on the Paseo (Passeig) de Gracia in Barcelona are often integrated with adjacent benches. Cast iron lamp posts by architects Pere Falques and Antoni Gaudi are considered important legacies of Barcelona's Art Nouveau heritage. (See Figure 9.3.)

The Lyon (France) 1989 lighting plan emphasized aesthetics, highlighting the illumination of 250 sites, including many historic buildings.[22] (See Figure 9.4.) This was later expanded by 50 additional installations. In 2004, the plan was revised, in part to reduce energy consumption and light pollution. The revised plan customized the treatment of different parts of the city. For example, riverbanks and some parks were designated as "calm zones" for lower illumination levels. Certain bridges with the best views of the city were called out for higher illumination. Architectural features and transportation facilities were also emphasized. The revised plan suggested custom treatments for different districts, determined in collaboration with local citizens. It also suggested the

▲ Figure 9.2

San Francisco "Path of Gold" Lamp Post. Lights are on Market Street, San Francisco's historic main street.

involvement of artists and designers in lighting planning. A process is underway to update the plan.[23]

9.3.2 Historic Residential Districts

The use of historic or ornamental lighting is popular in many residential districts, perceived by many as more attractive and distinctive than modern equipment. However, such decorative street lights can raise questions regarding safety, efficiency, and impacts. The City of San Jose, California responded to numerous requests for ornamental lighting within and outside historic neighborhoods by developing policies to support a mix of ornamental and standard street lights. City department managers stated: "The use of modern lighting fixtures may not be appropriate for areas of historic significance and it is not consistent with a City policy of preserving the look and feel of our historic neighborhoods."[24] Staff of the Planning/Building/Code Enforcement Department recommended that new lighting on residential streets in Historic Districts and Conservation

◀ Figure 9.3

Street Light Designed by Antoni Gaudi in Barcelona. Courtesy of Adobe Stock Photos. © dr_verner - stock. adobe.com.

◀ Figure 9.4

Lighting of Historic Sites in Lyon, France. Courtesy of Adobe Stock Photos. © vichie81 - stock.adobe.com.

Areas use ornamental fixtures with original illumination levels, typically lower than illumination levels for new installations. However, the Public Works staff recommended that intersections in such areas receive standard lighting, while mid-block locations could get new ornamental lights.

Historic lighting also may introduce higher levels of light pollution than comparable modern equipment. The popular acorn luminaire was often produced unshielded, potentially generating high levels of glare. However, there are LED options available with shielding and glare-reducing lenses.

Historic or ornamental lighting is popular in many residential and tourist districts, but raises concerns about lighting quality, energy use, and light pollution. While light-ing that appears futuristic can also be a branding tool, particularly for innovation districts and some tourist meccas, there are similar issues regarding light pollution.

9.4 Futuristic Lighting

Some cities use special lighting to project a futuristic image of a sophisticated and economically vibrant place. In such cases, the lighting is often not just for illumination but also intended to showcase Smart City capabilities and attract high-tech businesses and professionals. It can also be used as a tourist attrac-tion. Such lighting typically uses LEDs with dynamic elements that can vary colors or project images.

9.4.1 Innovation Districts

Numerous cities around the globe in recent years have developed "innovation districts." These are areas where cutting-edge institutions cluster near start-ups and business incubators, typically in fairly dense, transit-accessible, mixed-use districts.[25] Streets themselves in such districts are often viewed as living labora-tories, where innovative infrastructure like lighting can be tested. Good street lighting is also essential to supporting the robust pedestrian realm these districts usually champion.

Boston's Innovation District, one of the first in the U.S., attracted some 5,000 jobs to the 1,000-acre South Boston Waterfront. It was catalyzed by the completion of the Big Dig, undergrounding an arterial that had cut off the water-front from the rest of the city, as well as improved public transit. Attractions such as the world's largest startup company accelerator proved successful. In 2013, dynamic LED lighting was installed on the gateway to this innovation district, the historic Northern Avenue Bridge. The lighting project resulted from a partner-ship of non-profit Light Boston, the Boston Mayor's Office, and the Department of Public Works. The 1908 bridge is scheduled for demolition and replacement with a pedestrian/bicycle-only span after several years of closure.[26]

More recently, distinctive light attracted attention to the Lawn on D. Adjacent to the Boston Convention and Exhibition Center and the Innovation District, this public space offers games, interactive art, and public seating during warm months. Oval swings with solar LED lights have proven especially popular. The Lawn also has distinctive string lighting with custom poles. (See Figure 9.5.)

The Las Vegas, Nevada, Innovation District supports a range of innovative infrastructure including smart lighting, to attract businesses to relocate and as a staging ground for the Consumer Electronics Show. This is a highly influential annual trade show, whose January 2020 event attracted 171,000 attendees and

(a)  (b)

▲ Figure 9.5

The "Lawn on D" in Boston. Figure 9.5a shows bright string lights on a plaza, while Figure 9.5b shows the *Swing Time* installation of lighted oval swings popular with adults and children. Figure 9.5a courtesy of photographer Christian Phillips © 2021 and designer Sasaki Associates. Figure 9.5b courtesy of photographer John Horner © and designer Howeler + Yoon Architects.

6,500 media representatives.[27] Smart lighting has been promoted at the trade show and installed on streets outside the convention center.

Chattanooga (Tennessee), the first U.S. city with a citywide gigabit fiber-optic network, developed an innovation district as an alternative to big city tech districts. Along with the high-speed internet and relatively low cost of living, Chattanooga brands itself with distinctive lighting. For example, the Holmberg Pedestrian Bridge initially included uplit glass floor plates for a colorful after-dark feature.[28] Due to maintenance issues, the glass plates were replaced in 2020, but handrail lighting was added.

9.4.2 Tourist Meccas

Las Vegas Strip properties have been installing numerous ostentatious new and upgraded light attractions in recent years, updating a tradition dating back to gaudy neon light displays of the World War II era. The five-block canopy-covered Fremont Street Experience opened a $32 million LED lighting upgrade in 2020, featuring futuristic images and video seven times brighter than previously.[29] (See Figure 9.6.) Dubbed the world's largest video screen, 49 million LEDs and concert-quality music are used in Viva Vision.[30]

9.5 Artistic Lighting

Artistic lighting can take many forms. This section focuses on unique, personalized aesthetic expressions using illumination effects and/or equipment. Artistic outdoor lighting is rarely installed permanently in the street right-of-way. It is

> Artistic lighting is attractive and accessible to diverse audiences. It can change effects with the passage of time or inputs from passers-by or environmental data.

▲ Figure 9.6

Fremont Street Experience (Las Vegas) Canopy Light Show. Courtesy of Adobe Stock Photos. © Santi Rodriguez - stock.adobe.com.

more often sited in public plazas and parks. In such locations, however, it can also influence the pedestrian's street environment. Light as a temporal artform is often showcased in outdoor festivals, fairs, and expositions.

9.5.1 Public Art Displays: Static and Dynamic, Temporary and Permanent

Artistic lighting is a popular form of public art, often more accessible than a painting or sculpture tucked away in a museum. Artistic lighting can directly engage viewers, playing off the surroundings, while using dynamic effects at a scale greater than most museum and gallery pieces. Public art overall is a young, distinctive discipline, as noted by a proponent association: "Public art can express community values, enhance our environment, transform a landscape, or question our assumptions."[31] Light art as a sub-discipline has been steadily growing as light artists provide new perspectives on people's environmental experience, taking advantage of light's "symbolic, cultural, and psychophysical legacy."[32] Dozens of such artworks are showcased in an online compendium, ranging from installations on the Bow River in Calgary, Canada to internal illumination of hundreds of vacant houses in upstate New York to a September 11 commemoration with beams symbolizing the twin towers.[33]

Chris Burden's *Urban Light* is a sculptural assemblage of 202 historic street lights highly visible in front of the Los Angeles County Art Museum (LACMA). As a meta-artwork, it symbolizes the value of street lighting and other public infrastructure, breaking the boundaries between street objects and fine art. (See Figure 9.7.) As the late artist stated, a sophisticated society is one that is "safe

◀ Figure 9.7

Urban Light at the Los Angeles County Art Museum. Artwork © 2021 Chris Burden/ licensed by the Chris Burden Estate and Artists Rights Society (ARS), New York. Digital image © 2021 Museum Associates/LACMA. Licensed by Art Resource, NY.

after dark and beautiful to behold."[34] LACMA noted that the installation was not only "indisputably" its most popular work, but also often considered a symbol of the entire city, the second most populous in the U.S. The piece is accessible round the clock, and the lights were illuminated nightly pre-pandemic.

Temporal changes in light artworks and responsiveness to the environment are frequent distinguishing characteristics. Dynamic lighting can respond to sensors monitoring the immediate environment or to data pulled off the internet. Technologies for dynamic lighting include projection mapping, volumetric LED mapping, and structural video displays.[35] Projection mapping lights images onto surfaces like buildings. Volumetric LED mapping creates a "point cloud" in space. Structural video displays typically use special LED panels for video shows (which can be stacked like Lego blocks).

San Jose Illuminating Downtown

The City of San Jose, California, sponsored the Illuminating Downtown Program, which combines art and technology to help create a more "engaging" downtown district supporting the City's branding as the "Capital of Silicon Valley."[36] Promotional material specifically celebrates individual artists. Artworks include:

Sensing You (Artist: Dan Corson) – Painted circles and 81 illuminated rings were placed on the ceiling of the Highway 87 freeway underpass that change when activated by passing pedestrians and bicyclists. This artwork enlivens an important, but often overlooked space under an elevated structure, while recognizing the importance

▲ Figure 9.8

Freeway Underpass with Illuminated Rings Activated by Pedestrians and Bicyclists in San Jose, California. Light artwork *Sensing You* by Dan Corson, 2015. Artificially enhanced photograph. Photograph by Adrien Le Biavant. Image use courtesy of the City of San José Public Art Program.

of the non-auto users of the facility. (See Figure 9.8.)

Sensing Water (Artist: Dan Corson) – A freeway underpass was painted like flowing water. The mural is illuminated in evening hours at levels corresponding to current National Oceanic and Atmospheric Administration (NOAA) weather data.

San Carlos Lantern Relay (Artists: Steve Durie and Bruce Gardner) – Eight six-foot tall lanterns were placed near the downtown cultural and events district. Lanterns play a range of programs, responding to pedestrian flows. Pedestrians can activate the lanterns using buttons on poles supporting the lanterns.

Show Your Stripes (Artist: Jim Conti) – Pedestrians control illumination patterns on the façade of a downtown condominium building using phone codes.

The Stockholm (Sweden) installation *Colour by Numbers* uses lighting controllable by the public, either those walking by or online viewing a streaming webcam.[37] (See Figure 9.9.) Amateur artists can choose the floor on the upper portion of the 20-story tower to illuminate and also customize the color using an interface that allows different red, green, and blue mixes and varied intensities of each hue. The installation was a collaboration among artist Erik Krikortz, interaction designer Loove Broms, and architect Milo Laven.

Artistic lighting can relate to other arts. When the Nicollet Mall in Minneapolis reopened in 2017 after a multi-year renovation, it included 12 illuminated lanterns by artist Blessing Hancock featuring poems by local writers.

(See Figure 9.10.) The renovation also added 1,500 LED lights, including programmable lights on a two-block stretch called LightWalk.[38]

The Wabash Lights, in Chicago, is notable for combining illumination, placemaking, community participation, and education. The project's objective is to create "a place where art and technology come together in a canvas of light to foster public discourse and personal expression."[39] The project included extensive school outreach, using the artwork to promote student interest in Science, Technology, Engineering, Arts, and Mathematics (STEAM) education. The Lights are installed downtown on the underside of elevated train tracks. The project is an ongoing pilot with new interactive capabilities being added regularly.

Far simpler, but unique artistic effects can be achieved through additions to existing street lights. For example, Barcelona has used reflective metal mesh wraps on street light poles, and Quebec City, Canada has used giant lampshades.[40]

Artistic lighting can trigger concerns about light pollution, excessive expenditures, poor taste, or inappropriate themes. For example, a Vancouver, British Columbia outdoor chandelier was criticized for its $4.8 million cost, allegedly glorifying wealth and perhaps promoting gentrification.[41] The chandelier, by artist Rodney Graham, was installed as the required developer-funded public art component of a 59-story apartment tower. It drops and rotates twice nightly under the Granville Street Bridge over a roadway and sidewalk. The Mayor termed it "the most important piece of public art in the city's history."

9.5.2 Lighting Fairs and Shows

Lighting fairs and shows are scheduled events that provide elements of discovery and spectacle. They often mix lighting effects with music, public speakers, or other attractions. Lighting fairs are often held in the fall or winter months to attract tourists and residents. Some are major international extravaganzas, like "Vivid Sydney" and the Lyon Festival of Lights (whose popularity and economic impacts were discussed in Chapter 3). Lighting festivals have become a major, growing international tourism draw, with over a hundred events regularly held.[42] There is a global infrastructure of shared installations and logistical arrangements.

However, municipalities also hold more low-key promotions of light art. For example, Illuminate San Francisco is an annual light art festival held between Thanksgiving and January.[43] For 2019, the official San Francisco tourism agency mapped 12 installations for a self-guided walking tour (with a short ride on a historic streetcar). The path weaved through the downtown, civic center, and adjacent neighborhoods. It includes the Bay Lights and Day for Night projects described earlier. In 2020, the list of installations expanded to 33 permanent and eight temporary locations. Permanent installations included:

- Leo Villareal's *Point Cloud,* with over 28,000 LED bulbs rapidly changing color on a pedestrian bridge at the Moscone Convention Center
- Cliff Garten's *Monarch,* a stainless steel and LED sculpture of monarch butterflies, near a medical center plaza

◀ Figure 9.11

Programmable
Light Sculptures:
Vancouver Southeast
False Creek.
Light sculptures
were installed
for the 2010
Winter Olympics
for public light
shows. Designed
by DIALOG/PFS
Studio/CDM2
Lightworks. Photo
by Bob Matheson.

- Lara Haddad's and Tom Drugan's *Bayview Rising*, a 187-foot (57-meter) high mural on a grain elevator near a low-income neighborhood, with programmed colored lights making the mural seem animated.

Programmable lighting or light shows may be tied to special events. Vancouver (British Columbia) installed programmable light sculptures at Southeast False Creek to provide entertainment at a public plaza near the Olympic Village for the 2010 Winter Olympics. (See Figure 9.11.)

Lighting can be combined with different sensory inputs, such as the soundscapes used for 21 LED-illuminated vignettes (vistas, interactive elements, and culinary sites) at a temporary mile-long nighttime walk-through attraction on the grounds of the Fairchild Tropical Botanic Garden (Nightgarden Miami).[44] The show used fog effects that required sophisticated controls based on temperature, humidity, wind speed, and direction. Guests stayed 35 percent longer than the previous year due to the show. Lighting infrastructure was hidden as much as practical to support the park environment.

Illuminated drones can create mobile, dynamic lighted art, sometimes likened to fireworks shows. The drones are ultra-low-weight (under 330 grams) deployed in the hundreds under computer control.[45] Intel's LED Shooting Star™ system can create over four billion color combinations. While the drone shows are not directly linked to particular pedestrian routes, and can even be used for stadium shows, there is the potential for drone illumination to be used specifically to illuminate or energize walkways.

In a tribute to frontline medical workers during the COVID-19 pandemic, and in commemoration of the 75th anniversary of Dutch Liberation Day, the

Amsterdam-based artist duo "Drift" staged a performance titled "Franchise Freedom" in the skies above a Dutch hospital in May 2020.[46] The performance used 300 illuminated drones flying in the formation of a flock of starlings, before creating the symbol of a giant red heart that pulsed with light. The display concluded with the colors changing to those of the Dutch flag.

9.6 Lighting to Enhance Appreciation of the Natural Environment and Park Features

Lighting can enhance features of the natural environment and parks after dark. Different zones and activities can be contrasted, with illumination levels and color correlated temperatures (CCTs) appropriate to the type of activity. For example, a playground or historical feature will typically have higher illumination than vegetation. Lighting design should consider impacts on views, both on pathways into the park and within the park.

Toronto's Grange Park with the Art Gallery of Ontario Museum designed lighting to be generally unobtrusive to avoid detracting from trees.[47] Lighting designers worked with arborists to obtain uniform and safe illumination of paths without being blocked by the tree canopy. Also, designers avoided damaging the roots of trees. Uplighting illuminated and highlighted water features and an interactive playground.

Brooklyn's Domino Park lighting designers used cooler (higher CCT) LED lighting for highlight features for the five-acre waterfront park, such as walkways, play areas for bocce and volleyball, taco shop, playground, and more than 30 large-scale relics salvaged from the original Domino Sugar Factory.[48] Lighting Workshop of Brooklyn wanted to show the decay of the relics and to avoid distracting from views. The lighting designers limited color-changing LEDs to the water fountain.

Designers noted that warmer (shorter) wavelengths of light are perceived as being closer in proximity than cooler (bluer) wavelengths, and this was used in the layout. Walkways used continuous rail lighting at CCT of 3,000 Kelvin, while the upper portion of 21 steel columns was highlighted by 3,500 Kelvin up/downlight accent lights. Two original turquoise gantry cranes were lit by a mix of 4,000 Kelvin linear and point-source fixtures. Fixtures set within structural members of the crane emphasize its skeletal qualities while limiting glare. Ground-level illumination complies with the Department of City Planning and NYC Department of Parks and Recreation requirements for 1 foot-candle (fc) in walkable areas, with a maximum 10:1 maximum/minimum uniformity ratio. Marine-grade fixtures were used.

Lighting in parks should generally be understated to enhance the natural environment and views, but built features are often highlighted.

Lighting can also enhance the after-dark appreciation of water features, using reflections and ripples to vary lighting effects. For example, light artist Stephen Korns installed 78 battery-powered, blinking, amber-colored lights atop styrofoam squares in the Hudson River for a short period in 1980.[49] The lights were sited to interact with the shape of the adjacent Lower Manhattan piers and the ambient lighting.

One traditional method of bringing the natural environment to street lighting is to hang flower baskets or planters from street light poles. This has been used widely in cities such as Pittsburgh, Pennsylvania, and Victoria, British Columbia. Additional gardening required may be undertaken by a neighborhood or community association.

9.7 Lighting as Information

Banners or signs on street light poles have long been a simple method of providing information about events, organizations, places, and public service announcements. Lighting can also be used to project color codes, words, or symbols, typically onto buildings after dark. The scale of the projection can be

◀ Figure 9.12

Poem Projected onto Brooklyn Public Library. Poem supports imprisoned writers. Projection part of multi-city project by artistis Chemistry Creative. Courtesy of Chemistry Creative.

large enough to be seen for hundreds of feet or even blocks. This reaches large audiences who are not even seeking information.

The Empire State Building's tower lights are one of the most famous examples of color-coded displays on buildings.[50] Colors have changed as often as daily to honor dates or organizations since 1976. However, in 2012 a major LED system was installed to illuminate the upper floors and spire. For example, recently the tower was lit green and cyan in honor of Habitat for Humanity and sky blue with a rotating peace sign in John Lennon's memory.

The Shanghai Power Station's tall chimney hosts a giant LED-illuminated thermometer on the exterior of China's first state-run contemporary art museum.[51] It fits in with the extensive use of futuristic lighting, such as large LED screens on skyscrapers, that emphasizes Shanghai's status as a cosmopolitan city.

"The Writing on the Wall" project takes words of incarcerated poets and projects them onto the exterior walls of public buildings in the U.S. and Mexico.[52] The touring exhibit has visited cities, such as Detroit, New York, New Orleans, and Mexico City. Often the buildings used are part of the criminal justice system. (See Figure 9.12.)

Notes

1. Seb Braun, "How Paris Got Its Nickname, 'The City of Light'," Culture Trip website, 2019, https://theculturetrip.com/europe/france/paris/articles/real-reason-paris-called-city-lights/.
2. "LUCI – Lighting Urban Community International: About LUCI," LUCI website, accessed March 29, 2021, https://www.luciassociation.org/about-luci/.
3. Illuminating Engineering Society, *Lighting Practice LP-2 Quality Lighting Design for People in Outdoor Environments* (New York: IES, 2021).
4. Project for Public Spaces, "Lighting," Articles Search, accessed April 11, 2021, https://www.pps.org/search?query=Lights.
5. National Association of City Transportation Officials, *Urban Street Design Guide* (Washington, DC: Island Press, 2013); Jeff Speck, *Walkable City: How Downtown Can Save America, One Step at a Time* (New York: North Point Press, 2012).
6. Janet Lennox Moyer, *The Landscape Lighting Book*, 3rd ed. (Hoboken, NJ: Wiley, 2013); Dave Colangelo, *The Building as Screen: A History, Theory and Practice of Massive Media* (Amsterdam: Amsterdam University Press, 2020); Allan Jacobs, *Great Streets* (Cambridge, MA: MIT Press, 1993); and William H. White, *The Social Life of Small Urban Spaces* (Washington, DC: Conservation Foundation, 1980).
7. Jane Jacobs, *The Death and Life of Great American Cities* (New York: Vintage Books, 1961), p. 41.
8. Jan Gehl, *Cities for People* (London: Island Press, 2010), p. 180.
9. Jan Gehl and Lars Gemzoe, *Public Spaces, Public Life* (Copenhagen: The Danish Architectural Press, 2004).
10. Kevin Lynch, *Image of the City* (Cambridge, MA: MIT Press, 1960).
11. Nick Moser, One Hat One Hand Design, discussion with author, March 11, 2021.
12. Allan B. Jacobs, *Great Streets* (Cambridge, MA: MIT Press, 1993).
13. Philadelphia Center City District (CCD), "Lighting," CCD website, 2020, https://centercityphila.org/ccd-services/streetscape/lighting.

14. Project for Public Spaces, "What Is Placemaking?," 2007, https://www.pps.org/article/what-is-placemaking.
15. National Endowment for the Arts, "Our Town," website, accessed December 24, 2020, https://www.arts.gov/grants/our-town.
16. Feargus O'Sullivan, "Building Community through Better Street Lights," *Bloomberg CityLab*, October 4, 2017, https://www.bloomberg.com/news/articles/2017-10-04/cartagena-s-solution-for-better-street-lights.
17. "San Diego's Historic Gaslamp Quarter," San Diego Tourism Authority website, 2021, https://www.sandiego.org/articles/downtown/historic-gaslamp-quarter.aspx.
18. Michael Trimble, Executive Director, Gaslamp Quarter Association, Email, January 4, 2021.
19. Bruce Gellerman, "Beacon Hill's Iconic Gas Lamps Are Going Green," WBUR website, October 27, 2020, https://www.wbur.org/earthwhile/2020/10/27/beacon-hill-gas-lamps-leds.
20. Glen Korstrom, "Gastown Named World's Fourth Most Stylish Neighborhood," Business in Vancouver website, October 9, 2012, https://biv.com/article/2012/10/gastown-named-worlds-fourth-most-stylish-neighbour.
21. San Francisco Planning Dept., *Better Market Street Project: Draft Environmental Impact Report, Case No. 2014.0012E* (San Francisco: Planning Department, 2019), 4–8.
22. Jamie Bratt et al., *Best Practices in Placemaking through Illumination* (Blacksburg, VA: Virginia Tech Urban Affairs and Planning Program, 2010), https://www.arlingtoneconomicdevelopment.com/index.cfm?LinkServID=8B1450DD-D628-416F-9C7FB8702740004E&showMeta=0.
23. Frederic Durand, Manager, Urban Lighting Department, Lyon, email, December 10, 2020.
24. City of San Jose Public Works Department and Planning, Building, and Code Enforcement Department, Memo to City Council on Ornamental Street Lights, November 29, 2001.
25. Bruce Katz and Julie Wagner, *The Rise of Innovation Districts* (Washington, DC: Brookings Institution, 2014), https://www.brookings.edu/essay/rise-of-innovation-districts/.
26. "About the Northern Avenue Bridge Project," City of Boston project website, accessed February 11, 2021, https://www.northernavebridgebos.com/about.
27. Consumer Technology Association, *Attendance Audit Summary: Consumer Electronics Show 2020* (Arlington, VA: CTA, 2020), https://cdn.ces.tech/ces/media/pdfs/2020-ces-attendance-audit-summary.pdf.
28. Eryn Cooper, "Chattanooga's Holmberg Bridge Set to See $162 K Renovation," Newschannel 9 website, February 4, 2020, https://newschannel9.com/news/local/gallery/chattanoogas-holmberg-bridge-set-to-see-162k-renovation#photo-4.
29. Christopher Reynolds, "In Las Vegas, It's Lights, Cameras and More Lights," *Los Angeles Times*, January 28, 2020, https://www.latimes.com/travel/story/2020-01-28/vegas-bright-lights-displays.
30. "Viva Vision: World Largest Video Screen," 2020, https://vegasexperience.com/viva-vision-light-show/.
31. "What Is Public Art?," Association for Public Art website, adapted from Penny Balkin Bach *Public Art in Philadelphia* (Philadelphia: Temple University Press, 1992), accessed March 30, 2021, https://www.associationforpublicart.org/what-is-public-art/.
32. Russell P. Leslie, "The Evolving Face of Light Art," *Public Art Review* no. 30 (March 31, 2004).
33. "Light Art," Search, Americans for the Arts website, accessed March 30, 2021, https://www.americansforthearts.org/search/site/light%20art?page=1.

34. "The Story of Urban Light," *Unframed*, February 6, 2018, https://unframed.lacma. org/2018/02/06/story-urban-light.

35. Robb Pope and Nick Moser, "Digital Interactive Placemaking," IES Street and Area Lighting Conference, Dallas, Texas, October 28, 2020.

36. City of San Jose Office of Cultural Affairs, "Illuminating Downtown," City of San Jose website, accessed October 23, 2020, https://www.sanjoseca.gov/your-government/ departments/office-of-cultural-affairs/public-art/illuminating-downtown.

37. "Colour by Numbers," *Atlas Obscura*, accessed October 23, 2020, https://www. atlasobscura.com/places/colour-by-numbers.

38. "A Tour of the New Nicollet," Minneapolis Downtown Improvement District website, accessed November 21, 2020, http://www.onnicollet.com/design.

39. "Our Vision: An Interactive Public Arts Platform for Chicago," The Wabash Lights website, accessed March 30, 2021, http://www.thewabashlights.com/what.

40. City of Edmonton (Alberta, Canada), *Winter Design Guidelines: Transforming Edmonton into a Great Winter City* (Edmonton: City of Edmonton, 2016), https:// www.edmonton.ca/city_government/initiatives_innovation/winter-design.aspx.

41. Vancouver Sun, "Spinning Chandelier – Granville Street Bridge," November 28, 2019, Youtube, https://www.youtube.com/watch?v=2GKyTJ2jpuM; Brittany Orris and Katy Amon, City of Vancouver, Email, December 15, 2020.

42. Emanuele Giordano and Chin-Ee Ong, "Light Festivals, Policy Mobilities and Urban Tourism," *Tourism Geographies* 19, no. 5 (2017): 699–716, https://www.tandfonline. com/doi/full/10.1080/14616688.2017.1300936.

43. "12 Installations to See on the Illuminate SF Light Art Trail," San Francisco Travel website, November 27, 2019, https://www.sftravel.com/article/12-installations-see- illuminate-sf-light-art-trail; "Illuminate SF Festival of Light: Thanksgiving through January 23, 2021," Illuminate SF website, accessed January 20, 2021, https:// illuminatesf.com/.

44. "2020 IES Illumination Awards," *Lighting Design & Application* 50, no. 8 (August 2020), 33.

45. GrrlScientist, "Drone Light Shows 'Way Cooler' than Fireworks," *Forbes*, June 30, 2020, https://www.forbes.com/sites/grrlscientist/2020/06/30/drone-light-shows- way-cooler-than-fireworks/#1e5a6d5b42f1.

46. "Last Look: Flying Together," *Lighting Design & Application* 50, no. 7 (July 2020), 48.

47. Samantha Schwirck, "The Lighting Design for a Park in Downtown Toronto Succeeds by Disappearing," IES website, March 25, 2020. https://www.ies.org/lda/ good-neighbor/?utm_source=IES&utm_medium=Email&utm_campaign=Client%20 Updates&_zs=8K5NX&_zl=fjK32.

48. Samantha Schwirck, "The Sweet Spot," *Lighting Design & Application* 50, no. 1 (January 2020), 24.

49. Public Arts Fund, "Stephen Korns: Installation of Floating Lights in Hudson River," PAF website, 1980, https://www.publicartfund.org/exhibitions/view/installation-of- floating-lights-in-hudson-river/.

50. Empire State Building, "Tower Lights," accessed October 24, 2020, https://www. esbnyc.com/about/tower-lights.

51. Jenny Lin, "Shanghai: China's Bright New World? Dazzling Projections of Global Shanghai," in Sandy Isenstadt, Dietrich Naumann, and Margaret Petty, eds., *Cities of Light: Two Centuries of Urban Illumination* (London: Routledge, 2015), 115–122.

52. Jon Kalish, "'The Writing on the Wall' Finds Poetry Behind Bars, Projects It Onto Buildings," National Public Radio website, October 18, 2020, https://www.npr. org/2020/10/18/924338482/the-writing-on-the-wall-finds-poetry-behind-bars- projects-it-onto-buildings.

Chapter 10

The Future of Lighting for Pedestrians

10.1 Purpose and Scope of This Chapter

As described in the earlier chapters of this book, lighting for pedestrians has been changing rapidly and receiving additional attention in recent years. This is due to a confluence of factors, especially increasing pedestrian fatalities and injuries, rapidly evolving technology, and the involvement of a broad range of groups keenly interested in civic lighting decisions.

This final chapter suggests how these recent key trends may change in future years. It also identifies other trends that will likely influence both lighting design and pedestrian facilities planning and engineering over the next 5 to 20 years, including the graying of population, climate change, the rise of autonomous vehicles, other vehicle design changes, artificial intelligence and virtual/augmented reality, and transformative lighting technology changes.

Since a reliable crystal ball is not available to predict the future, it is wise to adopt a "future-proofing" approach. Future-proofing examines the key trends, shores up vulnerabilities, and tries to maximize flexibility.

This chapter closes with key ongoing research and potential new research directions. Links between research and practice are emphasized. Communication between lighting experts and other specialists can be enhanced by a common understanding of the issues and trends highlighted by this book.

10.2 Future-Proofing

"Future-proofing" is a valuable approach to minimize the risk of **obsolescence of technology**, such as smart lighting projects. Obsolescence can be physical (deterioration), functional (unable to perform the desired function), or aesthetic (out of fashion).[1]

Technology experts and futurists try to avoid the most obvious traps by carefully examining trends. They also build flexibility to adapt to future changes. For example, smart lighting poles can be designed to add or change out sensors and communication devices easily. But future-proofing can also be used by lighting specialists and others in developing local plans and projects. They can assess the potential risks and rewards of alternative technology paths.

 DOI: 10.4324/9781003149750-10

The process starts with examining technological trends and broad social and environmental trends, ranging from the impacts of climate change to evolving consumer tastes. One popular approach involves the assessment of **Strengths, Weaknesses, Opportunities, and Threats (SWOT)**. Strengths and Weaknesses are features of the technology or system, while Opportunities and Threats are external conditions that may influence its effectiveness.[2]

Alternative approaches to shoring up vulnerabilities and addressing external threats can be brainstormed and evaluated. Among **the tools to limit obsolescence** are:

- *Modular design* – design so that outdated components can easily be replaced while retaining the remainder of the device
- *Open architecture* – allow different companies and designers to modify and improve the design, rather than keeping the design details secret
- *Resilient design* – encourage hardened design that is less susceptible to shocks from climate change or other potential threats
- *Diversity promotion* – encourage alternative or competitive approaches, under the assumption that one or more will better survive regulatory barriers or other challenges
- *Life cycle costing* – consider the full costs of design, construction, operation, and maintenance including possible future changes to assumptions
- *Contingency plans* – develop plans for particular scenarios and implement when needed

The experience of the Segway personal transporter, as an alternative to walking, illustrates many of these SWOT points. The Segway was hyped initially as one of the greatest inventions of its time by Steve Jobs and other experts and launched with the *strengths* of enormous funding and marketing resources in 2001.[3] It responded to the *opportunity* offered by fairly short trips longer than easy walking distance. However, production was halted after less than two decades due to disappointing sales. Its demise was blamed on such *weakness* factors as lack of a broad market niche, failure to obtain objective feedback on the concept during development. and very high cost. It faced *threats* from cheaper, simpler competitors (like electric scooters) and regulators banning it from sidewalks and roads in many jurisdictions.

10.3 Key Current Trends

The key current trends that are making lighting for pedestrians such an important topic today are expected to continue to be significant for many years.

Pedestrian injuries and fatalities have been steadily rising in recent years, especially as a percentage of all traffic collisions. Some factors generally blamed for this rise are expected to continue supporting this unfortunate trend, such as driver and pedestrian distractions with smartphones, driver and pedestrian impairment, the popularity of larger motor vehicles, wide arterial street crossings, and other "pedestrian unfriendly" street design, plus limited police enforcement. The predominance of nighttime pedestrian fatalities will undoubtedly attract more attention from experts and advocates, who will focus on lighting as one potential response.

The COVID-19 pandemic's shift in shorter trips from public transit and shared rides to walking, bicycling, and driving will likely continue in the mid- to long-range even with vaccines and therapeutics, according to a panel of more than 60 transportation practitioners and academics assembled by the Pennsylvania Department of Transportation (PennDOT).[4] A significant share of travelers will likely be more sensitive to the possible health risks from close proximity to strangers in commuting. The pandemic's reduction in motor vehicle traffic also led some cities to reallocate street space to bicycle and pedestrian use or outdoor dining, which may outlive the pandemic. Increased use of electric scooters and shared, dockless bicycles is also highly likely, according to the PennDOT roundtable. This will be an increasing risk factor for pedestrians.

However, other factors will likely reduce pedestrian injuries. These include pedestrian collision avoidance systems and autonomous vehicles, changes in motor vehicle front-end design, headlight improvements, additional traffic calming measures and pedestrian safety devices, as well as stiffer laws against driving under the influence of alcohol or drugs.

Technology changes underway are very likely to accelerate. A U.S. Department of Energy (DOE) report forecasts that light-emitting diode (LED) lighting will reach 92 percent market penetration for street and area lighting by 2025 and 100 percent by 2035.[5] (See Figure 10.1.)

LED conversions are often a decision point for local governments or utilities to invest in smart lighting capabilities. The U.S. DOE report notes: "Of the LED lamps and luminaires installed in area and roadway applications, many are expected to be connected...The prospective capability to control and monitor each street light from one central location is highly appealing to municipalities and utilities alike."[6] By 2035, with current trends, nearly a third of area and street lighting luminaires are expected to be connected, compared to a DOE goal of 75 percent.

Smart city sensors housed on street lights are almost certain to increase. Applications such as traffic speed and volume monitoring are attractive to many

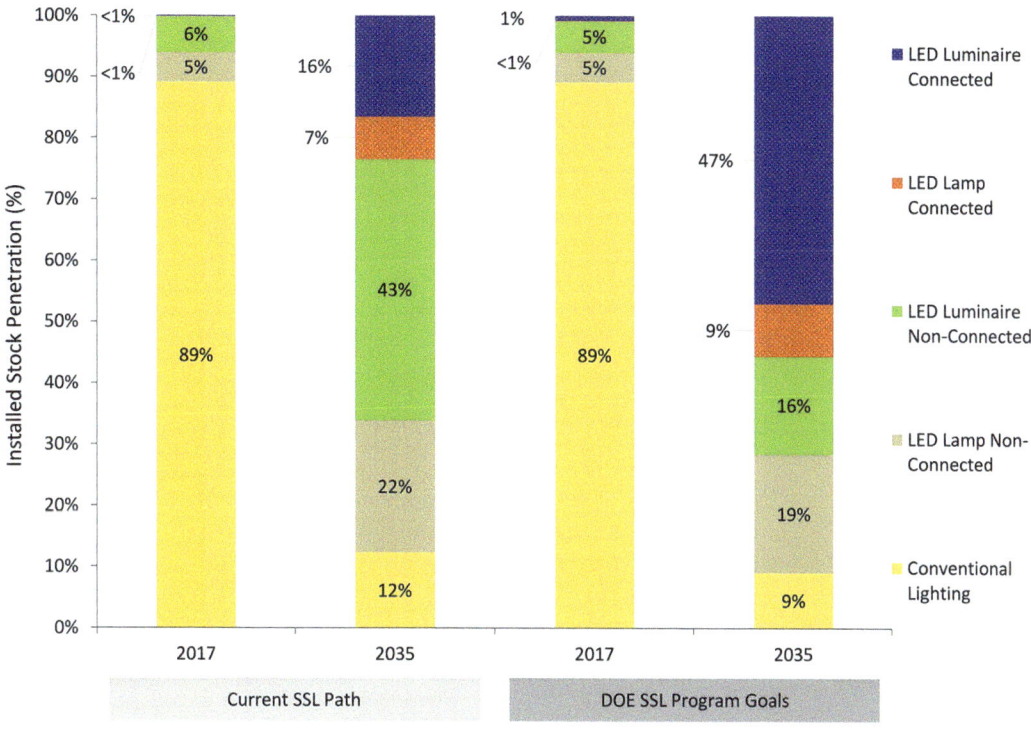

▲ Figure 10.1

Forecast LED Installed Stock for Area and Street Lighting. Figure courtesy of the US Department of Energy from Its Energy Savings Forecast of Solid-State Lighting in General Illumination Applications, December 2019 edition.

local governments. Their commercial potential can be seen by the entrance of major technology corporations into the field. Communication vendors like Verizon and AT&T tout their smart city applications. Business consulting firms, such as Deloitte, McKinsey, and PWC, churn out strategic analysis reports on the topic.

A wide range of stakeholders will likely be increasingly involved in lighting issues due to concerns about technology and pedestrian safety. There are many organizations concerned with pedestrian safety and the pedestrian environment, including those with a general concern for vulnerable groups like children and seniors. Privacy and cyber-security concerns around smart lighting are likely to stoke increased citizen involvement, particularly in an era of heightened skepticism about government and experts. Although light pollution is a central concern for the International Dark-Sky Association and some astronomy associations, it will also be an occasional concern for residents and wildlife protection groups. The health impacts of lighting will likely attract additional attention from medical groups and patient advocates. There may also be additional local advocacy groups like Light Boston, a nonprofit dedicated to encouraging and directly facilitating "aesthetically pleasing and environmentally-conscious" outdoor lighting, and the Flagstaff (Arizona) Dark Skies Coalition.[7]

10.4 Other Key Trends in the Long Term

Other significant trends that may affect lighting for pedestrians 5 to 20+ years into the future include social and technological changes:

- The graying of the population
- Climate change
- Autonomous vehicles
- Other vehicle design changes
- Artificial intelligence and virtual/augmented reality advances
- Transformative lighting technology changes

10.4.1 Graying of the Population

The average age in the U.S. and other developed nations has been increasing in recent decades due primarily to public health promotion, nutrition improvements, and medical advances. The U.S. Census Bureau projects that older adults 65 and over will outnumber children under 18 around the year 2034.[8] (In 2016, only 15.2 percent were 65 and older, compared to 22.8 percent under 18.) Japan already has greater than one in four people 65 years or older. Europe is also approaching that proportion of elderly.

Elderly drivers and pedestrians tend to have different lighting needs as their eyesight declines. Their overall health and mobility issues also make them vulnerable when walking. The elderly are more susceptible to glare due to increased scatter of light in the eye. The amount of light reaching the retina is reduced. There is generally reduced visual acuity (especially for peripheral vision), reduced contrast sensitivity, and reduced color discrimination.[9] With the decrease in light on the retina, disruptions in circadian rhythms are more likely. This can lead to problems sleeping, as well as other health problems. The elderly also suffer a higher proportion of eye maladies, such as macular degeneration, cataracts, and glaucoma.

There are specific lighting strategies to address such medical needs, such as reducing glare and selectively increasing illumination levels. Improving contrast is also helpful. Other measures to improve pedestrian safety generally become even more important for seniors, such as additional crossing time at signalized intersections, additional and improved curb ramps, more visible signs, and more stringent driver licensing procedures. Greater attention to fall prevention programs will also be valuable.

10.4.2 Climate Change

Climate change is spurring a massive push by governments, nonprofits, and many businesses to transform the generation and use of energy. This has led to the widespread adoption of LED technology. It also encourages more sparing use of lighting.

Increased use of LEDs in lighting is expected to substantially reduce electricity use. The U.S. Department of Energy (DOE) forecasts that the current path for adoption of LEDs and lighting control improvements will lead to savings of 62 quadrillion British Thermal Units (BTUs) through 2035.[10] However, if DOE program goals are met, the cumulative savings would be 78 quadrillion BTUs (equivalent to $890 billion in avoided energy costs). Adaptive lighting can further reduce electricity usage substantially. Street light poles can also be used for electric vehicle charging equipment, motor vehicle speed detection, and other measures that can address climate change.

Over the longer term, climate change may also lead to major infrastructure rebuilding and relocation, which will need accompanying investment in lighting. For example, waterfront areas in developed nations threatened by sea level rise will undergo major rebuilding with new or strengthened seawalls and relocated roadways and other infrastructure. Climate change will likely lead to migration, including a south-to-north movement within the U.S. (to escape higher summer temperatures), and perhaps the abandonment of some flood-prone or wildfire-prone locations.[11]

10.4.3 Autonomous Vehicles

Motor vehicle automation has been steadily increasing, with potential dramatic benefits for pedestrian safety. There has been extensive testing of autonomous vehicles by companies such as Waymo (a unit of Alphabet, Google's parent company) and General Motors (through its Cruise program).

Waymo started operating a fully **autonomous taxi service** open to the public in the suburban Phoenix metropolitan area in October 2020, following an extensive test period. During this test period, according to Waymo data, in 6.1 million vehicle miles (mostly with a highly trained driver ready to take over the vehicle if a crash seemed possible), no crashes resulted in severe or fatal injury, and there were no pedestrian-involved crashes in which the automated driving was considered at fault. Waymo vehicles operated day and night on public roadways with speed limits up to 45 MPH (72 KPH), although details were not available on the lighting or time-of-day conditions. Average human drivers logging the same number of miles would be expected to be involved in about five injury crashes.[12] But the Waymo evaluation stated that the number of miles driven was insufficient to draw statistically valid conclusions.

Waymo also plans to offer autonomous vehicle taxi service in San Francisco. However, it will be open only to approved "tester" passengers, with a safety back-up driver.[13] Cruise, the autonomous vehicle company backed by General Motors, has also been approved for self-driving taxis in San Francisco.

It is unclear when or if the U.S. motor vehicle fleet will turn over completely to fully autonomous vehicles. However, with a substantial share of vehicles typically on the road for well over a decade, a fully autonomous fleet seems highly unlikely before 2035.

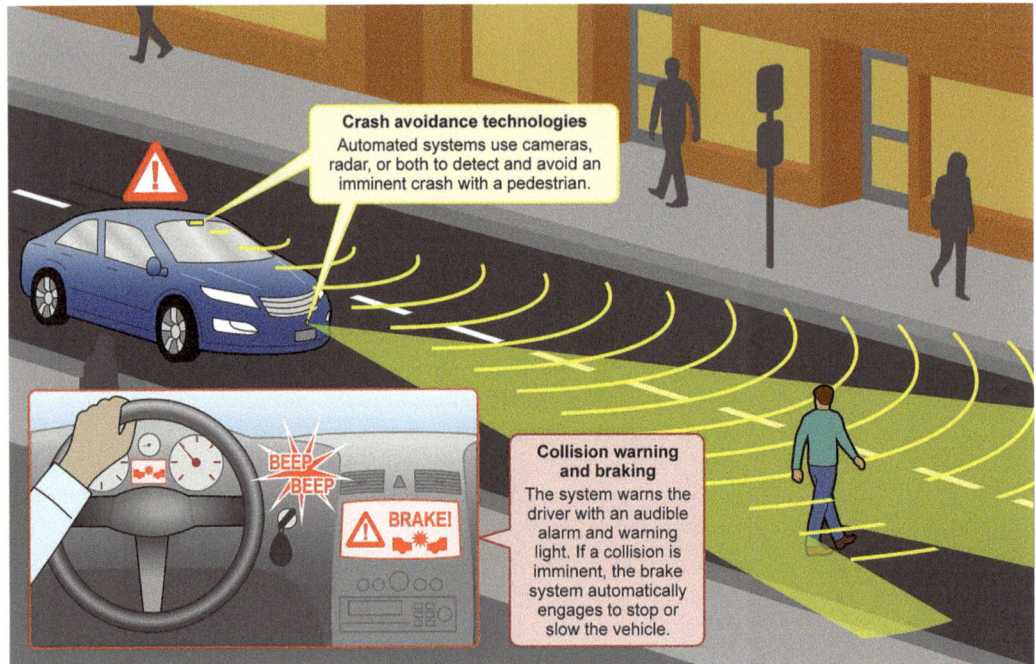

Source: GAO. | GAO-20-419

▲ Figure 10.2

Example of Crash Avoidance System. Courtesy of the US Government Accountability Office.

For pedestrian safety, automated **collision avoidance systems** short of full autonomy will still have benefits in the short term. (See Figure 10.2.) In particular, 56 percent of 2018 model year motor vehicles were equipped with Automatic Emergency Braking (AEB) with pedestrian detection as standard or optional equipment.[14]

Automakers representing 99 percent of the U.S. auto market agreed with the National Highway Traffic Safety Administration (NHTSA) in 2016 to make AEB standard on their new light-duty cars and pick-up trucks by September 1, 2022.[15] Heavier pick-up trucks up to 10,000 pounds should have AEB by September 1, 2025. AEB helps prevent or mitigate crashes, primarily rear-end vehicle crashes, based on Insurance Institute for Highway Safety track test standards. The agreement does not explicitly require pedestrian detection, but pedestrian collision avoidance systems are already being integrated into the AEB capabilities by

Autonomous vehicles include some sensors that do not depend on visible light, but they generally use cameras that benefit from street lighting. Pedestrian security and comfort will still be improved by street lighting. Scooter riders, bicyclists, and motorcyclists will also benefit from enhanced lighting.

many automakers. "NHTSA should build on this progress by ensuring that by 2025 all new vehicles come standard with more advanced systems that can detect pedestrians and work at highway speeds," stated Consumer Report's Vice President for Advocacy David Friedman.[16]

Collision avoidance systems and autonomous vehicles typically use a combination of sensors, some of which do not require light to operate. Cameras are currently helpful, for example, to perceive a new STOP sign most accurately.[17] However, RADAR (radio wave detection), LIDAR (laser detection), infrared sensors, and ultrasonic sonar do not require visible light, and other technological advances could also preclude the need for light.

When the AAA evaluated the pedestrian detection and automatic emergency braking systems on four different 2019 models, it found negligible **effectiveness at night** with only ambient light and low-beam headlights.[18] There was minimal if any braking for a pedestrian dummy and no notifications at all. The evaluation noted that owner's manuals generally warned the devices were not effective after dark or in adverse weather. It also noted that "cloudy or yellowed headlights" only produce 20 percent of the original light output. Performance during daytime was also of limited effectiveness. Under favorable conditions, with a dummy in a perpendicular crossing scenario, at 20 MPH (32 KPH) collisions were avoided 40 percent of the time, but at 30 MPH (48 KPH) three of the four test vehicles failed to reduce impact speed even by 5+ MPH.

The crash avoidance performance of test autonomous vehicles, such as Waymo reported above, is far better. However, the sensor systems of these autonomous vehicles are much more sophisticated and expensive than now offered to the general public on standard vehicles.

Those crash avoidance systems that use visible-light cameras in part **will benefit from street lighting in detecting pedestrians**.[19] Adaptive street lighting with occupancy sensors and/or crossing warning devices could even provide additional cues to vehicle sensing systems. Pedestrian-scale lighting may help with detection, especially if darker skin tones are harder to detect.[20]

Even if street lighting in 20 or so years is of limited value for motor vehicles to avoid collisions, pedestrians will still need lighting for security and reassurance purposes. Bicyclists, motorcyclists, and scooter/hoverboard riders will also benefit from street lighting. Pedestrians will still need to see sidewalk tripping hazards, signs, landmarks, and approaching strangers. Moreover, pedestrians will want to view approaching vehicles to select gaps for crossing at uncontrolled crossings and to react to an impending collision. It seems unlikely that most pedestrians in the foreseeable future would entrust their safety to a collision avoidance system on a stranger's motor vehicle after dark.

Also, the increasing use of sidewalks by **electric scooters**, which have minimal collision avoidance capabilities, may increase the need for sidewalk lighting. A U.S. Centers for Disease Control (CDC) study found an average of 14.3 injuries for 100,000 scooter trips.[21] Nearly three of five e-scooter injuries occurred on the sidewalk in a Washington, DC study.[22] Of these, about a third occurred where scooter use on the sidewalk is prohibited. The vast majority of injuries were to scooter riders themselves.

10.4.4 Other Vehicle Design Advances

Other key motor vehicle technology changes to benefit pedestrian safety include vehicle headlight advances, changes in front-end design, and improved testing of new cars.[23] **Adaptive headlight technology** used in Europe and other countries deploys advanced sensors and computing technology to increase roadway illumination and limit glare to oncoming vehicles. One automotive expert commented: "There is no doubt that within the next decade automotive lighting will have undergone a quantum shift in design and functionality."[24]

Headlamps can also be used to communicate with pedestrians. South Korean firm Hyundai Mobis has outfitted autos with headlamps that can detect a pedestrian up to 450 feet and project a red warning to the pedestrian.[25] When it is safe to proceed, headlamps can project a crosswalk symbol. Ford has proposed a similar concept for standardization for communicating with other drivers and pedestrians that the vehicle is stopping, starting, and accelerating.

Other vehicle **front-end design changes** have been studied or offered. These include a Volvo system in which the rear part of a flexible engine hood is raised if a front-end pedestrian collision is detected in order to protect the pedestrian from the engine block or windshield.[26] Other devices include soft bumper attachments, pedestrian protection airbags, and active front bumpers that automatically push pedestrians away from the engine block or windshield.

In 2015 NHTSA proposed **pedestrian safety tests** for its New Car Assessment Program, but it has not as of publication updated its program or provided milestones for implementing changes. NHTSA estimated that existing technologies could reduce annual pedestrian-involved crashes in the U.S. by 620 to 5,000 and reduce pedestrian fatalities by 110–810.[27]

10.4.5 AI, VR, and Transformative Lighting Technology

Smart lighting technology was described in detail in Chapter 6. Longer-term enhancements may include advances in **monitoring and communication with vehicles**, e.g., sensors could detect pedestrians crossing from behind a parked car and alert or stop approaching motor vehicles. **Automated pedestrian detection** already can provide additional crossing time for slower pedestrians, but an improved version could be integrated into traffic signals.

There are other ways that artificial intelligence and virtual/augmented reality may affect pedestrian safety or lighting. For example, **augmented reality**

Artificial intelligence and augmented/virtual reality will affect pedestrian safety and lighting in numerous ways beyond smart lighting, such as providing 3-D test experiences of lighting strategies for research and community involvement and providing "smart glasses" alternatives to improve low-light vision.

allows lighting designers to show stakeholders or researchers the effects of a proposed lighting scheme either in a conference room or on-site at 100 percent scale. (Augmented reality supplements or overlays sensory input from the "real world," while virtual reality completely replaces the "real world" with programmed sensory input.)

Smart glasses will likely improve low-light vision.[28] Current technologies use image intensification, active illumination, or thermal imaging for limited nighttime visibility improvements, but integrating more sophisticated infrared imaging can bring substantial improvement. This could help both fully sighted individuals and those suffering from night blindness.

Other long-term possibilities for lighting include: **pedestrian control of street lighting, personal drone lighting, and LiFi.** These options are more speculative than measures described in earlier sections. There is the theoretical potential for pedestrians to actively control illumination levels to suit their needs better using a smartphone or other device, communicating with smart lighting. A British insurer briefly tested prototype drones to light a path with LED flashlights.[29] Direct Line's Flight Series service used 15 quad copters, with one flashlight each, along with five hexacopters, equipped with three 200-watt lights. Perhaps one day personal drones could include video and emergency alert functions.

LiFi uses the visible light spectrum to transmit data through imperceptibly fast flickers.[30] As suggested in a 2011 TED Talk by Harald Haas of the University of Edinburgh, this could provide 100 times the bandwidth of WiFi with transmission speeds up to 50 MB per second. LiFi could be used to transmit road safety or traffic management data between street light pole sensors and motor vehicles or smart phones. However, LiFi requires direct line of sight, so, for example, the street light and smart phone or vehicle would need to be in an unobstructed line.

10.5 Ongoing Research Initiatives

Several valuable research projects were ongoing at publication time, and results were not available. This ranged from the collision impacts of LED lighting to visibility metrics and smart lighting evaluations.

Collision impacts of LED conversions were investigated by the Pacific Northwest National Lab.[31] Researchers reviewed the associations between a large-scale suburban LED conversion in the greater Philadelphia area and roadway crashes. The safety impact of color rendition, correlated color temperature, and output (lumens) of new luminaires were assessed. However, researchers did not have data on light levels at specific roadway locations, so it will need to be a fairly broad assessment of safety changes and will not separately address pedestrian-involved crashes.

Current visibility research by the Federal Highway Administration (FHWA) focuses on visibility along the roadway and of traffic control devices.[32] Assessing the cost-effectiveness of innovative markings, signs, signals, and lighting technology will rely on photometric and/or colorimetric measurements and human factors measures of effectiveness. FHWA also evaluates the safety impacts of roadway lighting and traffic control devices. It develops models of the visual information available to road users and the level of visual information required for appropriate performance. Metrics actively considered include color rendering, glare, and flicker.[33] Virginia Tech Transportation Institute is evaluating the need for supplemental illumination at midblock crosswalks equipped with Rectangular Rapid Flash Beacons or Pedestrian Hybrid Beacons.[34]

Smart lighting evaluations are being undertaken as part of major installations and pilot projects. For example, the City of West Hollywood (California) has initiated a six-month pilot program for "Smart City intelligent lighting technology and applications," coordinated with an LED conversion program.[35] Two vendors were selected to install "smart nodes" on luminaires. Eventual pilot evaluation criteria include:

- Ease of installation and maintenance
- Long-term projected costs/saving
- Quality and reliability of technology
- Security access controls
- New smart city features

Data from the smart nodes are expected to be used particularly to assist traffic and mobility studies, implementation of the City's Climate Action Plan, and refinement of traffic impact fees. [36] As part of the pilot program, the City developed Smart City Privacy Guidelines.

10.6 Desirable New Directions for Research

There are numerous valuable research possibilities to improve lighting for pedestrian safety and walkability. These efforts could further illuminate the benefits and adverse impacts of lighting, as well as compare actual practices in lighting design to guidelines and standards. The following are recommended topics.

10.6.1 Lighting Strategies and Technologies for Safety

Valuable research would include the safety impacts of **pedestrian-scale lighting**. Chapter 3 described the general safety benefits of lighting enhancements, including Crash Reduction Factors (CRFs) for motor vehicle collisions. However,

Lighting research that would be especially valuable for improving pedestrian safety and walkability should include expanding Crash Reduction Factor data on specific lighting strategies and naturalistic research on lighting and falls. Research can ultimately lead to: improved standards and metrics, reduced financial costs and adverse impacts, and effective education and community participation programs.

no CRFs were identified specifically for pedestrian-scale lighting. Naturalistic research would also be valuable regarding the actual impacts of pedestrian-scale lighting on pedestrian falls, considering different spacings or configurations.

Other research on **pedestrian falls** could focus on the impact of light levels and spectral content on frequency of sidewalk and crosswalk falls on actual streets.[37] The needs of pedestrians with mobility and visual impairments should be considered.

There are potential benefits to **illuminating street corners** and mid-block crosswalk approaches using lower-height lighting or supplemental illumination to make pedestrians more visible as they step into the crosswalk. This could be examined through photometric research, longitudinal crash evaluations, or driver test track experiments. Such research could also examine the effects of non-lighting variables such as parked vehicles, trees close to the intersection, pedestrian clothing, and the like.

While there are recommended light levels for separated walkways, which are often shared with bicycles, additional attention could be paid to comparing and meeting the lighting needs of **multi-use path users,** bicyclists, riders of electric scooters and other devices such as rollerblades and hoverboards.

Laboratory or test track research should consider additional **variables that affect the visibility** of pedestrians to drivers. The impacts of spectral content and glare on the required vertical illuminance would be a fruitful topic.[38]

10.6.2 Quantifying Benefits of Lighting beyond Safety

Chapter 3 presented a range of potential benefits for lighting beyond safety, including security and comfort, aesthetic enhancement, sense of place, information, and economic benefits. While the safety and crime reduction benefits have been quantified, other potential benefits are difficult to quantify. However, there are methods that could provide indirect data or rough estimates. For example, the **economic benefits** of historic or ornamental lighting to a commercial district may be approximated through visitor surveys in similar districts with and without special lighting.

Mood and mental health benefits have been researched for indoor lighting, but generally not for outdoor lighting. In part this is undoubtedly because the illumination levels considered therapeutic for Seasonal Affective Disorder and other conditions are far higher than achieved by most outdoor lighting.

Augmented and virtual reality will allow researchers to expose subjects to a range of realistic lighting conditions and observe their physical and psychological responses. Subjects can also report on their preferences.

Additional research on **factors affecting reassurance and facial recognition** beyond illuminance levels would be valuable. Researchers are interested in such variables as glare, spectral content, and uniformity.[39]

10.6.3 Costs and Adverse Impacts of Lighting

There are numerous opportunities for high-value research on financial costs and potential adverse impacts of lighting for pedestrians discussed in Chapter 4. For example, it would be valuable to have information on the **incremental life-cycle costs of pedestrian-scale lighting** added to a block or corridor already equipped with conventional roadway lighting.

Reducing light pollution will continue to be a major focus for lighting research. Glare is particularly challenging both to research and to minimize in practice. Discomfort glare is difficult to predict for a given luminaire, especially with the advent of LEDs. There are new tools that will aid in this work. High Dynamic Range Imaging (HDRi) allows measurement of luminances from multiple points, say on a non-uniform luminaire.[40] Devices to measure illuminance at and luminance within the eye are small enough to be worn easily in field tests.

10.6.4 Actual Practices in Street Lighting Design, Operations, and Maintenance

How closely municipal and utility lighting managers **adhere to lighting standards and guidelines** related to sidewalks and crosswalks would be useful information. It is also unclear how most transportation planners and engineers address lighting needs in their area and corridor improvement plans and projects aimed at improving pedestrian safety and walkability. The practical challenges both groups face in this area deserve investigation.

The **interaction of street lights and vehicle headlights** is not adequately addressed by simply following guidance for illumination levels, according to lighting human factors researcher Peter Boyce.[41] There are some cases "for which the combination of light distributions from the vehicle forward lighting and road lighting and the reflection properties of the target and road surface may lead to reduced visibility." Similarly, low visibility of pedestrians crossing at a specific intersection may be due to problems that are not confirmed by standard **intersection lighting metrics.** Basic recommendations for lighting at intersections do not quantify glare, contrast for pedestrians against pavement, or differences in illumination between street corner and crosswalk.

The potential benefits of tailored or customized **educational programs** for lighting and transportation staff to improve their ability to implement best practices could be evaluated. Similarly, strategies for public involvement and education for community stakeholders could be assessed. With increasingly complicated civic projects, such as smart lighting initiatives, there is a heightened need for government staff, policy boards, and others to absorb more technical and rapidly evolving information.

10.7 Improved Communication between Lighting Specialists and Others

This book has attempted to bridge an information gap between lighting specialists and others keenly interested in the subject but lacking the technical background needed to communicate effectively. There certainly is extensive sharing of information between lighting engineers and transportation professionals, especially traffic engineers. Some of the key reference works on roadway lighting were published by the Transportation Research Board, the American Association of State Highway and Transportation Officials, and the Transportation Association of Canada. The subject of this book was the focus of recent webinars sponsored by three transportation organizations. However, better mutual understanding of the procedures, concerns, and knowledge base of different professional and advocacy groups would be valuable.

This book was intended to provide a springboard to more productive communication and cooperation. Professional conferences and publications, project task forces, and informal communications are avenues to improve such information sharing, including addressing the research and practice questions raised in this chapter.

Notes

1. Brian Rich, "The Principles of Future-Proofing: A Broader Understanding of Resiliency in the Historic Built Environment," *Journal of Preservation Education and Research* 7 (October 14, 2016): 31–49.
2. Rich, "The Principles of Future-Proofing."
3. Paul Sloan, "A Lesson in Innovation: Why Did the Segway Fail?," Innovation Management website, May 2, 2012, https://innovationmanagement.se/2012/05/02/a-lesson-in-innovation-why-did-the-segway-fail/.
4. Yassmin Gramian, "COVID-19 Changed the World—Now We Must Ensure That Change Is for the Better," *Institute of Transportation Engineers (ITE) Journal* 91, no. 4 (April 2021): 45–49.
5. Navigant Consulting for U.S. Department of Energy, *Energy Savings Forecast of Solid-State Lighting in General Illumination Applications* (Washington, DC: US Department of Energy, 2019), 36, https://www.energy.gov/sites/prod/files/2019/12/f69/2019_ssl-energy-savings-forecast.pdf.
6. Navigant, *Energy Savings Forecast*, 66.
7. "Our Mission," Light Boston website, 2018, https://lightboston.org/?fbclid=IwAR0lIEBBztkeWAlufs_q5xzrJU_RM_SArIUkyj1T_R_xb8b0RSJmWJDGHX0; "To Celebrate, Promote and Protect the Glorious Dark Skies of Flagstaff and Northern Arizona," Flagstaff Dark Skies Coalition website, November 9, 2020, http://www.flagstaffdarkskies.org/.
8. Jonathan Vespa, "The Graying of America: More Older Adults than Kids by 2034," US Census Bureau website, October 8, 2019, https://www.census.gov/library/stories/2018/03/graying-america.html.

9. Peter R. Boyce, *Human Factors in Lighting* (Boca Raton, FL: CRC Press, 2014).

10. Navigant, *Energy Savings Forecast*, 2.

11. Abrahm Lustgarten, "How Climate Mitigation Will Reshape America," *New York Times Magazine*, September 15, 2020. https://www.nytimes.com/interactive/2020/09/15/magazine/climate-crisis-migration-america.html.

12. National Highway Traffic Safety Administration (NHTSA), "National Motor Vehicle Crash Statistics," NHTSA website, 2019, https://cdan.nhtsa.gov/tsftables/National%20Statistics.pdf.

13. Mark Bergen, "Waymo Brings Self-Driving Taxis to San Francisco—with a Catch," Bloomberg News, August 24, 2021, https://www.bloomberg.com/news/articles/2021-08-24/waymo-brings-self-driving-taxis-to-san-francisco-with-a-catch.

14. American Automobile Association (AAA), *Automatic Emergency Braking with Pedestrian Detection*, AAA website, October 2019, https://www.aaa.com/AAA/common/aar/files/Research-Report-Pedestrian-Detection.pdf.

15. Richard Read, "20 Automakers Promised to Make AEB Standard on New Cars," *The Car Connection*, March 17, 2016, https://www.thecarconnection.com/news/1102917_20-automakers-promise-to-make-automatic-braking-standard-by-2022; National Highway Traffic Safety Administration and Insurance Institute for Highway Safety, "Fact Sheet: Auto Industry Commitment to IIHS and NHTSA on Automatic Emergency Braking," NHTSA website, March 16, 2016, https://www.nhtsa.gov/sites/nhtsa.dot.gov/files/aeb_factsheet_031616.pdf.

16. Joe Young, "10 Automakers Fulfill Automatic Emergency Braking Pledge Ahead of Schedule," IIHS website, December 17, 2020, https://www.iihs.org/news/detail/10-automakers-fulfill-automatic-emergency-braking-pledge-ahead-of-schedule.

17. Andrew J. Hawkins, "Waymo's Next-Generation Self-Driving System Can See a Stop Sign 500 Meters Away," The Verge website, March 4, 2020, https://www.theverge.com/2020/3/4/21165014/waymo-fifth-generation-self-driving-radar-camera-lidar-jaguar-ipace; Alex Davies, "Waymo's Self-Driving Jaguars Arrive with New Homegrown Tech," Wired, March 4, 2020, https://www.wired.com/story/waymos-self-driving-jaguars-arrive-new-homegrown-tech/; Katie Burke, "How Does a Self-Driving Car See?," Nvidia blog, April 15, 2019, https://blogs.nvidia.com/blog/2019/04/15/how-does-a-self-driving-car-see/#:~:text=The%20three%20primary%20autonomous%20vehicle,as%20their%20three%2Ddimensional%20shape; Jordan Greene, "What's Best for Autonomous Cars: LiDAR vs. Radar vs. Cameras," Tech Briefs, September 11, 2020, https://www.techbriefs.com/component/content/article/tb/stories/blog/37699; Paul Tarricone, "In a World Without Drivers will There Be a Need for the Same Type of Roadway Lighting?," *Lighting Design + Application*, IES website, April 25, 2017, https://www.ies.org/lda/in-a-world-without-drivers/.

18. AAA, *Automatic Emergency Braking*, 2019.

19. Eleanor Leshner, Nico Boyd, and Alice Grossman, "Safeguarding Safety for Road Users Now While Planning for an Automated Future," *ITE Journal* 90, no. 4 (April 2020): 44–49.

20. Benjamin Wilson, Judy Hoffman, and Jamie Morgenstern, "Predictive Inequity in Object Detection," Cornell University arXiv open research website, February 21, 2019, https://arvix.org/pdf/1902.11097.pdf.

21. Laurel Moreno et al., "Characterization of Dockless Electric Scooter Related Injury Incidents in Austin, Texas, September – November 2018," 2019 Epidemic Intelligence Service Conference, https://www.cdc.gov/eis/downloads/eis-conference-2019-508.pdf#page=110.

22. Julia Cicchino, Paige Kulie, and Melissa McCarthy, "Severity of E-Scooter Rider Injuries Associated with Trip Characteristics," *Journal of Safety Research* 76 (February 2021): 256–261, https://trid.trb.org/view/1768298.

23. US Government Accountability Office, *Pedestrian Safety: NHTSA Needs to Decide Whether to Include Pedestrian Safety Tests in Its New Car Assessment Program*, GAO-20-419, GAO website, April 2020, https://www.gao.gov/assets/710/706348.pdf.

24. Peter Els, "Industry Spotlight: Automotive Lighting and Self-Driving Vehicle Safety," *Automotive IQ* website, August 29, 2019, https://www.automotive-iq.com/electrics-electronics/columns/column-automotive-lighting-and-self-driving-vehicle-safety.

25. "How Lighting Could Make Autonomous Vehicles Safer for Pedestrians," *Insurance Journal* website, January 14, 2019, https://www.insurancejournal.com/news/national/2019/01/14/514708.htm.

26. "Pedestrian Safety Systems," TLX Technologies website, 2020, https://www.tlxtech.com/markets/automotive/pedestrian-safety-systems.

27. GAO, *Pedestrian Safety*, 32.

28. Chunjia Hu and Guantao Zhai, "An Augmented-Reality Night Vision Enhancement Application for See-Through Glasses," *ResearchGate*, July 2015, https://www.researchgate.net/publication/283882292_An_Augmented-Reality_night_vision_enhancement_application_for_see-through_glasses/link/580a3f4f08ae2cb3a5d2ff5e/download.

29. "Fleet Lights Take Street Lights to the Skies," New Atlas website, November 17, 2016, https://newatlas.com/fleetlights-drone-streetlights/46499/.

30. Binary District Journal, "The Future of the Internet Is…Street Lights?," The Next Web website, August 3, 2019, https://thenextweb.com/syndication/2019/08/03/the-future-of-the-internet-is-streetlights/.

31. Jason Tuenge, "An Investigation of LED Street Lighting Conversions and Roadway Safety," IES Street and Area Lighting Conference, October 26, 2020.

32. FHWA, "Research & Development: Safety: Visibility," October 27, 2020, https://highways.dot.gov/research/research-programs/safety/visibility.

33. Boyce, *Human Factors in Lighting*, 595.

34. Ronald Gibbons, IES webinar on "Lighting and Roadway Safety," December 3, 2020.

35. City of West Hollywood Departments of Public Works and Economic Development, "City Council Consent Calendar Item: Smart City Intelligent Lighting Pilot Project," May 18, 2020, https://weho.granicus.com/MetaViewer.php?view_id=16&event_id=1192&meta_id=188015.

36. City of West Hollywood, "Agreement for Smart City Pilot Project Services with Ubicquia LLC," May 2021.

37. Steve Fotios et al., *CIE 236: Lighting for Pedestrians: A Summary of Empirical Data* (Vienna, Austria: International Commission on Illumination, 2019), doi: 10.25039/TR.236.2019.

38. Fotios et al., *CIE 236*.

39. Fotios et al., *CIE 236*.

40. Boyce, *Human Factors in Lighting*, 602.

41. Boyce, *Human Factors in Lighting*, 415.

Index

Note: **Bold** page numbers refer to tables; *italic* page numbers refer to figures.